Praises for

Touché

As I read *Touché* by author Todd Griffee, I could not help but think of the quote by Mark Twain: "The two most important days in your life are the day you are born and the day you find out why." Griffee captures the essence of this profound statement through his personal journeys and experiences. He finds his own purpose through the unique sum of all parts which appeal to all ages and life experience levels. A well-worthy read.

—**Chriss Lyon**, author, *A Killing in Capone's Playground: The True Story of the Hunt for the Most Dangerous Man Alive*

***

Todd Griffee is a two-time All-American fencer at Notre Dame who got there through hard work, determination, and the help of several extraordinary people. His roadmap is a helpful guide on how to persevere and succeed in life. What makes this book so wonderful is that through teaching the reader his path to success, he reveals himself – plainly, directly, and honestly. It's a real gift to glimpse the soul of an author, and Todd's book gives you just that.

—**Gia Kvaratskhelia,** head coach, fencing, *Notre Dame Athletics*

***

I worked with Todd Griffee for several years prior to my retiring. The first thing I recognized about Griffee was his work ethic, and as I got to know him, he shared with me some of his experiences at Notre Dame. Reading *Touché*, put all of those achievements into perspective. Griffee always had the motto, "If I can help, I will help," and that motto lives up to itself with Griffee laying out the method for his success in a format to help others achieve their goals. Whether you are a Notre Dame fan, alumni, or subway alumni like myself, you will enjoy this glimpse into what it means to be a Notre Dame athlete.

—Special Agent James McCartney, NCIS (ret)

***

Todd Griffee's *Touché* is a compelling narrative of resilience, faith, and determination. As a two-time All-American fencer at Notre Dame, Todd's journey is a living testament to the power of hard work and unwavering commitment. Each chapter is a blend of personal stories, practical lessons, and engaging activities, making it a valuable guide for anyone seeking to overcome obstacles and achieve their dreams.

Having had the privilege of being Todd's cello instructor for several years, I've witnessed firsthand his remarkable discipline and dedication. His passion for excellence was evident in his musical pursuits and mirrored in his athletic and academic achievements.

Todd's story is not just about fencing; it's about a holistic approach to life where discipline, love, and faith converge to create a path to success. Todd's ability to turn challenges into opportunities and his relentless pursuit of his goals serve as a powerful example for readers of all ages. His candid storytelling and practical advice not only make this book an inspirational read, but also a practical roadmap for those looking to achieve their own greatness.

Whether you are a young adult seeking direction, a professional in need of motivation, or simply a fan of Notre Dame, Todd's story will inspire you to pursue your dreams with faith, determination, and love.

—**Audrey Jo Williams,** private strings instructor at *Music with Audrey LLC,* co-founder & co-director of *New Horizons | Bloomington IN,*_member of *Friday Musicale,* member of *Arts Alliance of Greater Bloomington,* associate member of the *Suzuki Association of Indiana*

***

Touché illustrates the enormity of finding your calling and the inimitable value of a great mentor. The book lays out a blueprint of how to succeed at your vocation. Author Todd Griffee writes from experience, having reached the pinnacle of success as an All-American fencer for Notre Dame University.

—**James A. Watkins,** author

I had the absolute pleasure to serve alongside Todd at a naval base just three or so hours south of Notre Dame. He was a quiet, unassuming, and highly-skilled law enforcement professional. I can attest that he does not pay lip-service to the words he writes in these pages. He lived and breathed hard work, perseverance, tenacity, teamwork, and humility day in and day out.

In *Touché*, he charts the path to success and fulfillment in an accessible, relatable, and engaging way. A great read from a great American!

—**Tim "Petey" Powers**, commander, *United States Navy (ret)*

I had the privilege of working with Todd Griffee for a few years as his cello teacher. Todd was an excellent student, and to the extent I was able to teach him the cello, he taught me about discipline and one's work ethic through his diligent preparation and indefatigable attitude. *Touché* is sure to impart these and more of Todd's lessons to readers.

—**Dr. Kevin Flynn**, visiting assistant professor of cello at *Oklahoma State University*

Touché

A Notre Dame All-American Fencer's Spiritual Guide to Success

by

Todd J. Griffee

Published by KHARIS PUBLISHING, an imprint of
KHARIS MEDIA LLC.

Copyright © 2024 Todd J. Griffee

ISBN-13: 978-1-63746-269-0

ISBN-10: 1-63746-269-7

Library of Congress Control Number: 2024946495

Cover Photo Courtesy of Todd Griffee

All KHARIS PUBLISHING products are available at special
quantity discounts for bulk purchase for sales promotions,
premiums, fund-raising, and educational needs. For details,
contact:

Kharis Media LLC
Tel: 1-630-909-3405
support@kharispublishing.com
www.kharispublishing.com

I would like to dedicate this book to my mother, Leslie Griffee (1945-2009), who was always my biggest cheerleader. I would also like to dedicate it to my father, Joe Griffee (1943-Present), who was always there to support my dreams, and Coach Michael DeCicco (1927-2013), who took a chance on a local kid and taught him priceless lessons, which would last a life time.

CONTENTS

CHAPTER SUMMARY

1) Introduction – The purpose of the book is to serve as a roadmap for children, young adults, and adults. Every chapter has a lesson, story, and activity. There is importance in writing things down because they become real and official when you do. I cover some basics of fencing and share my hopes for you.

2) Chapter 1 – Nothing's Impossible. I was a young child bending spoons because, like other children, we believed anything was possible. When adults limit children's worlds, it becomes harder for them to dream big. My father always said I could do anything, and my mum was my biggest cheerleader. Even though you might not know your purpose, you need to open yourself up to the possibilities and positive energy. Write "I Believe!" on a sticky note and put it on your bathroom mirror.

3) Chapter 2 – Know your History – This chapter describes my ancestry, my family's history, and our love of Notre Dame. It is important to know your own history because it provides some insight into your uniqueness.

4) Chapter 3 – Foreshadowing – This chapter tells the story of Paul Hornung (1935-2001; UND graduate;

Green Bay Packer 1957-1966) rubbing my pregnant mother's stomach at a party at Ave Maria Press so I would be a Notre Dame All-American. It encourages you to look for clues from God.

5) Chapter 4 – What is my Purpose – At ten years old, I devised a plan on how to discover my purpose, which was to become a knight. This translated to a federal agent in modern times. Be quiet and connect with God. Make a list of possible purposes.

6) Chapter 5 – Make a Plan. My plan was to go to Notre Dame, be an athlete, join the service, and obtain a career in federal law enforcement. Even though plans change, it is important to keep to the task. You must be flexible and be prepared to change your Plan A to Plan B, to X, Y, and Z, if necessary. It is imperative to write down your plan.

7) Chapter 6 – Remove the Excuses – My cousin's fencing really stuck. It was simple…a way to incorporate the old with the modern. But how would I get there? We had no computers, cell phones, internet, etc. I used pen and paper, the yellow pages, and started calling, making contacts, and gathering information. After many failed attempts, I finally spoke with Coach Mike DeCicco at Notre Dame. Identify obstacles, and come up with solutions.

8) Chapter 7 – Execute the Plan - How would I find fencing lessons? When I finally talked to Coach DeCicco, he thought I was in college, and invited me

to train with the varsity team. Write down how you have reacted to opportunities, and visualize success. Think three positive thoughts for every negative thought which creeps into your mind.

9) Chapter 8 – Do What it Takes to Make it Happen – Fencing equipment was expensive so I had to work odd jobs, and ride my bike to practice when I could not get a ride. Training required extra effort like running four miles after practice. Getting good grades to get into college was also important. Make a list of requirements for achieving your goal.

10) Chapter 9 – Mentors and Supporters – My mum and her friend sold raffle tickets at a bowling alley so I could take the trip to California to compete in Junior Olympic National Championships, where I placed 5th in the country. Make a list of mentors and supporters.

11) Chapter 10 – Confidence and Proficiency– I scored on the world fencing champion at 16. The chapter discusses the importance of being confident. Identify methods of gaining proficiency.

12) Chapter 11 – Lessons in Sportsmanship and Life Lessons – Coach told me his "crossing a stream" metaphor and the importance of giving an opponent the best you have. I needed to learn to lose with grace, but let it burn and train harder. Coach DeCicco made it a point to lift up others. It is imperative to learn from your mistakes, and humbly accept constructive criticism. Coach DeCicco told me to walk around with

a blade in my hand all the time, so I did while walking to practice. Write down lessons that you have learned from mentors.

13) Chapter 12 – Decisions and Sacrifices –I was forced to give up high school football, soccer, and boxing. It sounded harsh, but you often need to make sacrifices to complete your goal. How bad do you want it? Make a list of things that interfere with accomplishing your goal.

14) Chapter 13 – How do we Pay for This? – At that time, Notre Dame did not have fencing scholarships, but I was accepted to ND. I consulted with Chris, the team captain, about Navy ROTC. Although I had a dismal Navy ROTC interview, I had a successful Army ROTC interview, which fit perfectly with my original plan to join the service. I worked 20 hours a week to help with tuition. Look ahead at potential problems, write them down, and write down appropriate solutions.

15) Chapter 14 – Freshman Year – I had an incredible workload. Notre Dame was extremely hard, time was in short demand because of ROTC, Rangers, work, class, practice, after practice studies, getting acclimated, and travel. I worked throughout my summers. The importance of tutoring: Notre Dame is no joke. You must work for your grades. Organize your schedule.

16) Chapter 15 – Sophomore Year Burn-out – I was burned out and wanted to drop out. Dr. Koellner

shared some valuable words of advice: Nobody will ever be able to take your Notre Dame degree from you. Contact mentors before making rash decisions.

17) Chapter 16 – Balance – I discovered meditation and balance. I learned to focus on mind, body, spirit, music, and utilized imagery to hone my skills. I established a prayer and routine to work smarter, not harder. However, hard work was still required. Explore meditation to increase your focus, organization, and to hone your skillset.

18) Chapter 17 – The Weak Links – Coach's comments in *The Observer* called the epee squad the "weak links" so I had t-shirts made to use as motivational tool to prove everyone wrong. The epee squad became the "strongest link." I ended up winning my first All-American Award, and was elected captain for my junior year. Turn obstacles into opportunities.

19) Chapter 18 – Defend Your Honor- An opponent school captain was being disrespectful of "Our Lady," and won the bout, which followed. Trash talking was not my thing. Write your core beliefs. What would you fight for?

20) Chapter 19 – Take Responsibility for Your Actions - The ridiculous assistant coaches were making my life miserable. I made a mistake by going out drinking, and was fired as captain. Despite utilizing the humility lessons I had learned earlier, and taking responsibility for my actions, I was eating humble pie the week before

NCAA National Championships. I was feeling low and like a failure. When you are feeling low, make a list of everything for which you are grateful.

21) Chapter 20 – Leave it all out on the Strip – Despite all of the distractions, I was able to get the team to a 2nd place finish by competing in the championship match. I completed this task with a .500 record, and won the First Team All-American Award, with the support of my family, friends, and God. Write down prepared scenarios for troubleshooting, and know you can pick yourself up when you are knocked down.

22) Chapter 21 – Life on Life's Terms – Senior year was a great season, but I had a terrible outing at the NCAA National Championship. At the end of the year, I was commissioned as an officer in the Army, and graduated from Notre Dame. I was confident I contributed as much to Notre Dame as Notre Dame had given to me. Notre Dame will always have a special place in my heart. Reflect, and list lessons learned.

23) Closing – Coach DeCicco gave me a huge compliment, stating that despite being like a bull in china factory, I refused to lose. I share principles which I utilized throughout my adult life. *Our Lady of Victory, Pray for us.*

INTRODUCTION

St. John Southworth, Westminster Cathedral, London, UK
(courtesy of Todd Griffee)

I know what you are thinking: "Another rah-rah book about Notre Dame." Another story of a small town underdog who makes good. Another *Rudy©*. Honestly yes, of course people enjoy Notre Dame rah-rah stories, especially when they involve an underdog, an unlikely small town kid who makes good!

What makes this book different is that I break down how I did it, including spiritual and life-learning lessons. I've included an instructional template others can utilize to meet their goals. Despite this book telling the story of my Notre Dame fencing career, the lessons I learned and discuss can apply to any goal or activity.

Having raised three independent children, having a schoolteacher for a wife, and having to work with young sailors, I have noticed a huge difference among young adults today, as opposed to when I grew up.

We live with an education system that is test-orientated around math (not the same math either), and English (without cursive and phonics). History, art, geography, music, and social studies are but mere footnotes in their curriculum. Providing the tools for children to use their imaginations, understand lessons from the past, appreciate creativity, and to understand basic human interactions has been lost. The Covid-19 epidemic did not help either. These younger generations deserve the ability to dream, make goals, execute plans, and troubleshoot obstacles. They deserve the opportunity to have passion inspire them, and to experience the sense of adventure in the journey toward a life goal.

It is my goal to share my experiences in a manner that reveals "school of hard knocks" failures, along with spiritual revelations that enabled me to meet my dreams. The lessons I have learned built a solid foundation of trouble-shooting, which served me well in the hard journey of life. In one of the *Rocky*© movies, Rocky tells his son that life is not about how hard you can hit; it is about how hard you can get hit, and keep moving forward. The speech is spot on, and boils down to one lesson: The only defeat occurs when YOU give up. I have always put it in my mind that I will refuse to lose. As General George S. Patton (1885-1945) exclaimed, "I will fulfill my destiny!"

The lessons I learned on this journey have served me well throughout my adult life. My life has not been perfect, but I had tools, not only to negotiate my troubles, but also to thrive through them. I have been through abuse, deaths, divorces, job loss, financial devastation several times, prosecution, and rejection. Despite all of the challenges, I live a very happy, meaningful, and successful life, and I am truly blessed. It is my hope that if you can see how God worked miracles in my life, you can have faith He will work miracles in yours!

In each chapter, I provide a lesson with some suggestions as to how to apply those lessons. I STRONGLY encourage the reader to write down their thoughts on each activity on the pages provided. Something magical happens when you write things down. My goal is not only to create hope and faith with an inspirational journey, but also to provide some hard-learned nuggets of truth. These are mere suggestions of what worked for me. Every individual is so unique, so feel free to adapt them to what works for you.

Let us cover the elephant in the room—religion. I grew up Roman Catholic, and attended Catholic elementary and high schools. Quite honestly, I enjoyed the structure, and loved most of my teachers. My experience with Catholic education was a very positive one. I also enjoyed the theology classes and environment at Notre Dame. Later in life, due to a divorce, and re-marriage, my Catholic affiliation was troubled. Not because of the Catholic Church's teachings, but because of my situation. If I could not abide by their rules, I needed to find membership within an organization where I could fully participate.

To decide where I wanted to go, I looked through my ancestral history. My grandmother, Jane Morton, whose mother was a Southworth, gave me a family tree with extensive information. She charged me with carrying on our English history. I learned that King Edward I, who fought in the First Crusade (1096-1099), was my 21st great-grandfather. I had another ancestor, St. John Southworth, who was martyred for being a Catholic priest during the Reformation. I also had a relative, Sir Thomas Southworth, who became the first ancestor to convert to being a Protestant.

After prayer and meditation, I found the Church of England to be a proper fit for me. I did not have to compromise any beliefs I grew up with in the Roman Church, but could be a member in full participation. This book is less about religion, and more about the spiritual journey of building a friendship with the Higher Power of your choosing. For me, that is the Father, Son, and Holy Spirit of traditional teaching.

Your religion, God, or Higher Power is your business. The goal is to recognize there is something more powerful than yourself. You need to become teachable, and learn the tedious, mundane, and small tasks you are given. What is important is to humble yourself, and open yourself up to what God has in store for you. God wants us to happy, joyful, and free, but there is a lot of work and lesson learning to get there. Happiness is not a goal in itself; it is a byproduct of doing the right thing in front of you.

Now a bit about fencing. Everyone thinks fencing is cool due to movies. Swashbuckling movies about Zorro, and the Three Musketeers were always a hit. However, fencing is not a popular sport because it is too fast to make a good spectator sport on television. Literally, 1/20th of a second is the difference in time between you hitting your opponent for a point or your opponent hitting you for a point.

Facts About Fencing: The Three Weapons and Scoring

The foil

There are three weapons involved in fencing. The first is a foil, which was never a true fighting sword. It has always been a practice or sporting sword. It has a small bell guard with a very flexible, square shaped blade. Hits are only scored using the tip. There is a button on the tip that, when compressed, closes the circuit and scores the point. The only target you can score on is the torso, excluding the head and arms. Foil fencers wear a metallic lamé (vest), which scores hits on target. The foil also has "right of way", which means if you extend your blade to attack your opponent, your opponent has to parry your blade to gain the right of way to counterattack and score. It is a very confusing

21

system that relies on the discretion of a judge. In other words, it is a good way to get cheated. Foil is a very fast and exciting sport to see in person.

The sabre

The sabre is a fencing sword that derived from the cavalry sabre. The bell guard wraps around the entire hand, and the blade is usually T-shaped. Hits can be scored with the tip, the top 1/3rd of the blade, and the entire cutting edge. Sabre fencers wear lamés like foil fencers, but they cover from the waist up, to include the head and arms. Sabre is a vicious sport where you can get some serious welts. It is fast and also exciting to watch, but it also has the "right of way" rules that make it easy to get cheated by a director.

The epee

Epee was my weapon of choice. The epee is a French dueling sword, and the only change to the weapon—since the 18th Century—is the electronic tip at the point. The epee is the heaviest of the fencing weapons, with a large bell guard, and stiff, V-shaped blade. If you removed the tip, you could easily run your opponent through in a duel.

I have seen blades break and fencers get run through the leg. It was gruesome. I also had a blade break and hit me in the ribs (after I first scored the game winning point), and my ballistic jacket stopped it from running me through, but it cracked a couple of ribs. The amount of weight it takes to compress the electronic tip to complete the circuit is the same amount of weight it would take to break the skin to draw blood.

Epee was the last weapon to become a sport because it was used for actual dueling. The purpose of the duel was not necessarily to kill your opponent, but to draw first blood. There are no "right of way" rules or target restrictions. You can hit your opponent in the face, foot, legs, body, but the favored target was the hand because it is the opponent's target closest to you. It takes a good eye and a lot of target practice to make a hand shot. It was also the easiest target of opportunity to keep your distance from your opponent so you were not hit. Simultaneous hits were also scored.

Epee is a bit more slow and deliberate game to watch, but is easier to follow (less likely to be cheated) because there are no rules. "Hit or get hit" is the mantra. Fencing is a very physically, and mentally challenging sport. You are literally fighting your opponent, which really gets your adrenaline and emotions fired up. The challenge is to focus that energy, and utilize the mental strategies required to win. Fencing is an art and is like playing Chess on a physical level.

Even though fencing is a lot more obscure than football, Notre Dame fencing is legendary. My coach was Michael DeCicco, whose teams won five National Championships, and he was the most winningest coach in Notre Dame history. Coach DeCicco was a powerhouse to say the least. When I was 18, he was in his 50's, and when he would give me a lesson, he would run me into the ground. I thought my legs and arm would fall off.

It is also funny to note that when my grandmother, Jane, was in her fifties during the same period, she would kick my ass in tennis and table tennis. The woman had Yoda© like reflexes, and was in excellent shape. Here I am, a nationally-ranked epeeist in my age group, and I'm being destroyed by these people in their fifties! It brings me hope now that I am in my fifties.

Coach DeCicco was a fiery-tempered Italian, and he was gifted in bringing out the best in his athletes. His fiery temper and his drive to win made him the best fit for being my mentor because he understood my nasty temper and my drive to win. He taught me so many valuable life lessons that have helped me throughout my life to this day. He was also the director of the Academic Advising Program at Notre Dame. His program is the reason Notre Dame's graduation level amongst athletes has been so high.

During my tenure, Notre Dame had a 100 % graduation rate for athletes. The program did not mean the classes were easy (the classes were bloody hard), or that someone else did your work. *Au contraire!* If your grades dropped, you were assigned to tutoring, and could be sitting in a study group with a Heisman trophy winner and many other football players who would end up playing with pro teams on Sundays. Nobody did our work for us, gave us the answers for tests, or wrote our papers for us. The bottom line is that Notre Dame made you do the work.

I have a lot of respect for Notre Dame's integrity. The lessons I learned on this journey from being a child discovering his future, to becoming a Notre Dame All-American fencer became the foundation for all of the adventures and trials I encountered on my journey as an adult. Someone once told me a foolish man does not learn from his mistakes, a smart man learns from his mistakes, but a wise man learns from the mistakes of others. It is my hope I can provide some focus to others so they might learn from my mistakes.

Later, in Chapter 17, I discuss how Coach DeCicco referred to the epee squad as the "weak links" of the fencing team to the school newspaper, *The Observer*. I used the lessons learned on this journey to transform the epee squad into the "strongest link." Utilizing the lessons of my journey, I want to give every reader the opportunity to be the "strongest link" in their organization, family, endeavor, team, or in their quest. As I've said, my desire is to reveal how God performed miracles in my life, and to bring hope that He will work miracles in your lives.

I structured this book to act as a roadmap for attaining your goals. Each chapter has a lesson, a part of my journey which demonstrates the lesson, and ideas for implementing the lesson. These things have worked for me, but feel free to adjust them to what works for you. My goal is to inspire people to understand EVERYONE has the potential to accomplish great things.

1

NOTHING'S IMPOSSIBLE

Jay's Lounge, Niles, MI (courtesy of Joe Griffee)

Every great story starts with humble beginnings. I grew up under the shadow of the Golden Dome in the small town of Niles, MI, just 10 miles away. Niles was an interesting place to grow up. I remember swimming in the St. Joseph River off Island Park as a wee one. That was right across from where the old Fort St. Joseph was located. What remains now is underwater due the installation of a dam. The fort was built by the French, and was taken by the British after the French and Indian War (1754-1763). In 1783, after the Treaty of Paris, which ended the Revolution, the Americans claimed the fort. Finally, the Spanish took it for a couple of days during the Spanish American War (**1898**). Niles is therefore dubbed the city of four flags—a lot of cool history for a small town in the Midwest.

My mother, Leslie, was a server at various restaurants, and eventually worked at the University Club at Notre Dame. She was a dynamo. She maintained a great household, even though she worked many double shifts. She was the planner for all the family events, and family was a very big deal. My father was a frogman in the Navy, and was a "plank owner" on the USS Von Steuben. He was big into sky diving, and was a part owner in a tavern, Jay's Lounge, which his father created back in the early 1960's. He later became successful selling x-ray equipment to hospitals in the region.

I learned my sense of adventure and my "Protestant" work ethic, as he would call it, from him. He was Protestant, and later converted to being Roman Catholic for my mother, who was a big time church goer. We were not rich, but my parents worked hard, not only to provide necessities, but also to create the fundamental foundations required to succeed in life.

A distant family member abused me at an early age, and the only reason I mention it is because it helped to shape who I was to become. When a child is abused, there are a few possible paths to choose. You either become an abuser and continue the cycle of abuse, you continue being abused because you never learned what personal freedom or choice mean, or you become the person who will be damned if they let this happen to anyone else. I made an early decision to be the latter, and was determined that I was not going to be a victim again. The calling to be a defender of the innocent was instilled, and served me throughout my law enforcement career.

So often I have asked myself, *why did I have to suffer this?* People like to blame God, but God did not do this. An evil person did. Because of that clarity of attributing responsibility, I have a different level of empathy and compassion for others. Later, in law enforcement, I could actually tell a victim, "I know how you feel, "and mean it. It is also important to note that many people see themselves as damaged goods; they then use that as an excuse not to succeed. Have faith, because God enjoys using broken people to complete extraordinary tasks to show His hand in the achievement.

Enough of the mood killer. My childhood was wonderful. I am writing this around Christmas, and am reminded of all the wonderful family times spent over the holidays. We always had tons of family come for the holidays, and it was exciting to see my aunts, uncles, and cousins. These were magical times with sledding, sleigh rides, Christmas caroling, and table tennis tournaments. Even though my grandparents and mother have passed away, I am eternally grateful for the brilliant holidays they provided.

On weekends, when I was five, my father would take me to Jay's Lounge before they opened for the day. At that time, Jay's was a typical Michigan tavern. On the wall there was a giant moose head that Grandpa Jay (Bumpa) shot in Canada. The Griffees were plasterers, so there was a sign on the wall that said, "Get plastered with Griffee." The food was incredible.

I had chores to do. I would clean the parking lot, sweep the inside, and stock the coolers. I made a couple of extra bucks that I would use to buy Godzilla™ and dinosaur models. After, my chores, my Grandpa Jay would cook me breakfast. I can still recall the taste of the sunny side up eggs, bacon, and hash browns, with jellied toast. While I ate my breakfast, I would watch TV—usually, *Voyage to the Bottom of the Sea*©—but one day there was a magician on TV. He was bending spoons just by gently rubbing them. I thought to myself, *wow that is really cool.* I went to the kitchen, grabbed a spoon, and started rubbing it. I remember focusing my concentration on it, like the guy on TV, and you know what happened? I bent the bloody spoon! Not once. Not twice. I could do it every time, at will. The spoon would just feel hot and it would just bend over as if I hit it with a blowtorch.

I remember showing people at Jay's, and they thought it was a trick, but it was not. I was bending spoons like a Jedi™ long before anyone knew what a Jedi™ was, but one day somebody said, "Hey, kid, you cannot do that. That's impossible." It was as if someone hit a switch. I was never able to bend a spoon again. Even though I knew I could do it, the magic was turned off.

Children have this innate manner of thinking outside the realm of what is possible, and they lose that when grownups limit their understanding of the universe. I think that is why Jesus told His apostles to be child-like. Not because children are simple-minded like sheep, but because they believe without seeing. Children have a faith very different from Doubting Thomas, who had to touch Christ's wounds to believe. Children do not limit their potential. Be open to the miracles. How many children have had their abilities stunted because some jackass said, "That's impossible"?

Fortunately for me, my parents always believed I could do anything I put my mind to. I remember loving dinosaurs, and learning all their scientific names. My parents supported that interest, and encouraged my enthusiasm. My father would tell me that I could be the President, an astronaut, or a paleontologist, as long as I put in the hard work to achieve it.

My mother was always my biggest cheerleader. I still remember her cheering for me when I was fencing, "Like a cat Todd, like a cat." In her mind, the way I fenced was like a big cat pouncing on its prey. The greatest gift a parent can give their child is to instill the core belief that nothing is impossible! My favorite motto is that of the British Special Air Service (SAS):

"Who Dares Wins!" It is simple, yet so true. Believe you can accomplish the impossible and you will. If you have children, please encourage them to believe!

There is something very special about the positive energy released into the cosmos, when you feel you can do anything. When you put out positive vibes, they tend to return to you. However, also remember faith without action is dead. I recall I was told a long time ago about a man who was stuck in a flood. He was swimming and prayed to God for help. A few minutes later, a man in a canoe stopped by and asked the man to get into the boat, but the man replied, "No, because the Lord will provide." Shortly after, a motor boat came by and the operator asked the man if he needed help. The man said, "No, the Lord will provide." Twenty minutes later, a helicopter flew over to rescue the man, and once again, he replied, "No, the Lord will provide." The man drowned. He went to heaven, where he met God. The man asked God, "Why did You not save me? You taught me you would provide." God replied, "I sent two boats and a helicopter. What more could you ask for?"

The anecdote is humorous, but points out some real truths of how God works in our lives. Faith and our willingness to take action by doing the next right thing in front of us is key. And yes, I mean "the next *right* thing in front of us," not the next thing right in front of us. It is important to take action with the conviction you will reach your objective.

The assignment for this chapter is to write your ideas down. There is a special thing that happens when you put pen to paper, whether you are writing down ideas for attaining your goals or writing down what troubles you so you can identify ways to improve. Writing it down helps get it out of your head, and into action. It is the first step in making it real. Being in government service, the slogan is, "If it is not in writing, it did not happen."

There is something very therapeutic and exciting about writing it down. My trouble was ideas would bounce around in my head, and I was always worried I would forget those ideas. Writing the ideas down not only helped me retain those ideas, but also to organize them. I would also worry about things, and by writing them down, it seemed to make it better. I was no longer dealing with the Boogie Man; I was dealing with a problem on paper. I learned early on that procrastination only added to my anxiety. I cannot stand things hanging over my head. I made it a regular routine to write down what was bothering me, and write out possible solutions. Then I would pray about it and turn it over to God. The next day, I would look at my list and determine if there was anything I could do to accomplish the possible solutions. It might be writing a letter, making a phone call, or doing some research. Once I finished what I could accomplish for that day, I would pray about it and turn it over to God. I would continue the process until the problem was solved. This routine helped to keep my mind from racing and accomplishing nothing. Try writing down your ideas. Even if is just a sticky note on your bathroom room mirror that says, "I believe!"

2

KNOW YOUR HISTORY

Southworth Coat of Arms, Samlesbury Hall, UK
(courtesy of Todd Griffee)

History has fascinated me for as long as I can remember. Whether it was dinosaurs, knights, Redcoats, WWI fighters, WWII fighters, generals, aviation, or classical music, it was all fascinating. I always had a strong connection with England, and one day my Grandma Jane explained why. She told me I had blue blood running through my veins because our family came from royalty. She gave me a packet with our family tree and a ton of information. She charged me with carrying on our family history. I found out I was the 21st great-grandson of King Edward I, and was related to a long line of knights, sheriffs of Lancashire, members of Parliament, and members of Elizabeth I's privy council. History became real because I was related to these people.

The Southworth family lived in Samlesbury Hall near Preston, England. When I finally was able to visit the hall, the connection was extremely powerful, like a punch to the gut. The sense of déjà vu was inexplicable. I remember seeing a very crooked short wall and touching it. I said out loud that the wall was crooked as hell; it felt like it was not the first time I touched that wall, and I said that. The only other time I felt déjà vu that powerful was the first time I put on my Redcoat captain's uniform for a reenactment.

My coworkers at NCIS are always teasing me about my age because I am so much older than most of them. So, one day, I asked if they wanted to see a picture of me when I was in the army? I showed them a picture of me in my Redcoat uniform and said that was back when men were steel and ships were wood. It was a good laugh for everyone. One thing I have learned in life is not to take yourself too seriously.

The last time I saw my Grandma Jane alive, she reminded me of the blood that ran in my veins, and told me the story of when her father had to leave her mother for a while. He told her, "You have the blood of kings. You can be strong and handle yourself." She insisted I continue the family history.

My grandfather was Richard Putz, and he grew up in South Bend, IN. He started going to all the Notre Dame home games from his childhood. I believe he missed one home game in 40 years due to my cousin Martha's wedding. I still do not think my cousin has lived it down getting married during a Notre Dame home game. We always have a good chuckle about it tailgating every year. All of his children became huge Notre Dame fans, which flowed to all of the grandchildren. I was always very close to my grandfather and mowed his yard for years. It is because of him that the love for Notre Dame was burned into my psyche.

My father was a frogman in the Navy, and was quite the skydiving daredevil. My Uncle Jeff played softball for the Navy, and my Uncle Ed was in Military Intelligence in the Army. My step-grandfather was in the Army Air Corps in England during WWII. His stories of fighter pilots flying through the Eiffel Tower, and bombers making it back to base with holes the size of a Jeep® always fascinated me. These role models laid my foundation of service. It amazes me that during WWII approximately 9% of Americans served in the military. Today less than 1% serve in the military.

The Frogman, Joe Griffee, USN (courtesy of Joe Griffee)

This is my history. No better and no worse than anyone else's history, just unique to me. This history was been a source of strength for me through the years. Just as no one can take away my Notre Dame degree, no one can take the blood from my veins. I have also learned a lot about myself through studying my ancestors. It has helped me understand my very nature. It is important for everyone to understand where he or she came from. Regardless of race, religion, culture, and nationality, embrace your roots and be proud of them. It provides the building blocks and raw material you have to work with. This book is designed for readers to celebrate their uniqueness, and express it during their journeys.

Activity

The activity for this chapter is to do some research into your roots. Talk to the elder people in your families. With the passing of every generation, volumes of information are lost. Write down family traits and interesting details of your ancestors, and see if they provide insight into yourself.

_____

_____

_____

_____

_____

_____

_____

_____

_____

_____

_____

_____

_____

_____

_____

_____

_____

_____

_____

_____

_____

_____

_____

_____

3

FORESHADOWING

Paul Hornung Poster, Notre Dame Stadium,
(courtesy of Bonnie Griffee)

I explained what a hard-core Notre Dame fan my
Grandpa Richard was. Well, so were his friends. He
was friends with Maxine and Ken Hall, who worked at
Ave Maria Press at Notre Dame. He was also friends

with Fr. John Reedy, who was the editor and publisher of the *Ave Maria©* magazine that the Ave Maria Press produced. This group of friends would have parties during home game weekends.

My mother attended one of these parties when she was pregnant with me. Also at the party was 1956 Heisman Trophy winner, and All-American Paul Hornung. Legend has it, according to my mother, she had Paul rub her belly so that I would be a Notre Dame All-American.

When my mother recounted this story to me, it planted the seed of my being a Notre Dame All-American. Even as early as the second grade I just knew I was going to Notre Dame and would be an All-American athlete.

I think God has a sense of humor, and enjoys leaving little clues of our destiny. Sometimes you do not see them in the present, but they make sense when you look backwards in time. Just listen to the human-interest stories while watching college or professional football. They are full of anecdotes celebrating this very thing.

Activity
The activity for this chapter is to ask your family about interesting tidbits from your childhood, and write them down.

4

WHAT IS MY PURPOSE?

"Nessie Hunting," Loch Ness, (courtesy of Todd Griffee)

Now that we have covered the foundation and raw material we are dealing with, it is time to dream!

As I was writing this in my home office/cello studio, memorabilia from my childhood surrounded me. Above my door is a wooden plaque my mother gave me when I just a wee one. It reads, "Todd's Office." Seeing it takes me, back to the memories of

43

that little office—barely a walk-in closet or pantry—from my childhood. The room was filled with dinosaurs, a Godzilla™ monster, and airplane models. I can still imagine the Fokker triplane, Albatross biplane, F4U Corsairs, and loads of other fighter aircraft. There were books ranging from King Arthur, to the Three Musketeers, to mysteries, to books about General Patton, to the Loch Ness Monster, to paleontology, to aircraft. There were sketches I drew of dinosaurs and aircraft. It was in that world where I can remember sitting at my desk at approximately 10 years old, and wondering what I wanted to be when I grew up. There were so many interesting possibilities.

The more I thought about it, the more important the question became. I felt like I needed to start planning, but I had no direction. I remember becoming somewhat distraught, almost obsessed with finding my direction. I thought about becoming a fighter pilot, and that remained something feasible for future military service, but I was unsure of doing it for a lifetime. The next option was to be a paleontologist or cryptozoologist (study of missing animals). The thought of exploring remote areas of the globe in search of animals thought to be extinct for millions of years sounded appealing, but how would I support a career like that financially? Becoming a paleontologist would require eight years of college, and probably would not be a great return on the investment. It did not leave a lot of room to be financially able to raise a family.

Nothing seemed to stick, so I continued to pray. Then one day, as I sat quietly at my desk, the thought came crashing in: *Stop limiting your question to this timeframe!* I asked myself, *if you could be anything you wanted to be at any time in history, what would you be?* The answer was immediate. A knight! I would be a knight! In a world of "might makes right," knights used their skills and courage to utilize the codes of chivalry to protect the weak and innocent. They used their "might for right." The Japanese equivalent to the knight was the samurai, which translates to "servant."

My next step was to figure out how I could be a knight in modern times. That answer was just as immediate! A police officer or federal agent. Those are the modern-day equivalents to being a knight, right down to the shield on the chest, and putting one's life in harm's way to protect the weak and innocent. I remember calling my grandfather with the exciting news. After much deliberation, I figured out what I wanted to be when I grew up: a G-man.

I love history, and at this point in my life, I was reading a lot about General George S. Patton. One thing I found interesting was his statement that it was important to be proficient in the use of all weapons, whether they were obsolete or not. Patton was a sabre fencer, and competed in the 1912 Olympics in Pentathlon (shooting, epee fencing, running, swimming, and horseback riding). I took that advice to heart.

My grandfather was quite the archer, and he gave me a 50-pound recurve bow. I learned how to shoot it well. The secret was not only practice, but also figuring out what worked and what did not. I would focus on how it felt when I shot a bullseye, then reproduce those mechanics and concentration with each proceeding shot. This developed follow-through and muscle memory. This practice system has served me well throughout my life, whether it was fencing, shooting, darts, or mastering the cello. The formula is the same.

When I was 13 years old, my cousin Andy started at the University of Notre Dame to study architecture. During his freshman year, he was required to take a physical education class, and he chose fencing. He told me all about fencing, and I was instantly intrigued. This was another profound revelation. Fencing was a way to bridge the desire to be a knight and apply that training into a modern framework. My mind was made up. I was going to fence!

At some time in their lives, everyone is completely lost; they're stuck trying to decide where God wants them. This can range from schools, majors, careers, relationships, and so forth. Somehow, as a 10-year-old child, I was miraculously able to find the solution.

The secret was to be still, be quiet, and listen to your inner voice. I prayed and listened. The other component was writing down ideas. I wrote out thoughts about all the careers I considered, with all of their pros and cons. When I was quiet and still, I would imagine myself doing that job. I never tried to force the ideas.

It was like trying on shoes. You do not force them on to make them fit. You pick the correct size, walk in them, check them for comfort, determine if they will meet their intended purpose and, of course, look in the mirror to see how they look. Some shoes are just utilitarian, but you will know when you find the perfect shoe. That is the feeling when you start to find your purpose. You feel like you were meant to do this.

Activity

The activity for this chapter is to find time to relax in a quiet spot, and connect with your Higher Power, as you understand them. For me, the traditional God fits. Find what fits for you. There are tons of meditation and relaxation exercises online which might be of further assistance. Let your mind go and do not force your thoughts. Start with asking God what He has in store for you, and then the meditation is the listening part. Do not be discouraged if it takes a while, but rest assured it will work. The second part of this exercise is writing down your thoughts after each session, and then thinking them through, and making more notes. Check those shoes to see if they fit. When you find the right direction, you will have an epiphany, and know it is correct.

Todd Griffee

5

MAKE A PLAN

"Young LT", 1993, Fort Custer, MI.
(courtesy of 822 MP CO)

As young Army cadets and officers, it was burnt into our psyches to make a plan. During exercises, the instructors would scream in our ears, "Lieutenant, an artillery strike just killed half your platoon, what are you going to do? OOOPs, too slow. Three more soldiers are dead. DO YOU HAVE A PLAN YOUNG LIEUTENANT?"

I felt sorry for the others who did not have the habit of making plans. I had a salty Sergeant First Class tell me early on, it was better to have a bad plan than no plan at all. Without a plan, you are just adrift, being acted upon by every force in the universe and without direction. I would always be flooded with tons of ideas flowing into my brain spontaneously; sometimes I felt like my brain was a hamster running on a wheel. When I learned the art of formulating plans, it gave order and focus to those random ideas. It became even more useful when I wrote down my plan. It freed up my mind to continue brainstorming or concentrate on the task at hand.

After being enlightened with becoming a modern day knight, I had to make a plan. I talked to a friend of my father who was a DEA agent, and received loads of useful information. My quest to be a federal agent would require a bachelor's degree. I also wrote the Secret Service, and asked what I should major in at college. They responded that sociology would be a desired major. I kept, filed, and referred to that information often. Let me remind you I was only 13 years old when I started gathering career information and even set up a filing cabinet with information. Never tell yourself it is too early to research and plan. It is important to take action to formulate your plan.

My next plan was where I would go to college. That decision was a no-brainer—Notre Dame. I kept it in the back of my mind. The US Naval Academy could be a backup plan, if I were not accepted into Notre Dame. Honestly, I focused all of my attention on Notre Dame. Normally I would not encourage someone to put all of their eggs in one basket, but this was a burning passion which could not be extinguished.

Now I had to figure out how military service would fit into my plan. I was intrigued by my father's stories of his captain calmly reading the latest James Bond novel, while the submarine's crew listened pensively to the deafening screech of the ice above and around them scraping the ship's hull beneath the North Pole. I wanted to be a part of something like that. Even though this plan was not solidified, I wrote it on my notepad to figure out later. The brilliant thing about writing things down is that you can keep adding to your notes.

That leads to fencing. I did not know how fencing was going to fit in the plan. I just knew it would. Looking back, I am shocked by how it was the very core thing which made everything else possible.

When you make a plan, it does not have to be complete or rigid. It needs to be fluid and adaptable. The key is writing it down and getting it started. In life, rarely is a plan executed with Plan A. You usually need to go to Plan B, Plan C, and so on. In Miyamoto Mushashi's (1584-1645) *The Book of Five Rings* (1645), he provides insights to the martial art of Kenjutsu (styles of Japanese swordsmanship). These insights are often applied to many disciplines other than sword fighting. The concept which stands out is being fluid like water.

Water is powerful enough to cut rock, yet pliable to flow around obstacles. Your plans must be like water. They must be powerful, yet fluid enough to continue progressing. You have to start somewhere, and write it down so it is not random ideas bouncing around in your skull.

Activity

The activity for this chapter is to write down your goals, plans, and how you will execute them. The only important thing is to get started, and write down those ideas. It will keep evolving and changing as you progress toward your accomplishment. Proverbs 16:3 states to commit your work to the Lord, and your plans will be established.

_____

_____

_____

_____

_____

_____

_____

_____

_____

_____

_____

6

REMOVE THE EXCUSES

Cousin Andy, Stan, and Uncle Jeff (courtesy of Todd Griffee)

After being an Army officer, supervisor, business owner, and parent, it seems like I spent countless hours hearing: "I cannot find it," "I cannot do this," "This will never work," "This is too hard," "Nobody else is doing this," "I do not have the right tools,"— the list of excuses still goes on and on. People love to hide behind excuses because it allows them to explain or justify being lazy. However, there is a more dangerous reason people hide behind excuses: Fear. People hate change – good or bad - they hate it. It means there is more work to be done, and the future is full of unknowns.

In all my endeavors and roles, it appeared most of my energy was spent removing other people's fears, and eliminating their excuses. When I was teaching my daughter Megan to play piano, the first words out of her mouth would always be, "I can't do this." It drove me crazy. I would have to constantly reassure and remind her anything was possible. She could do this. It would be during those times when I would revisit how I learned to overcome obstacles at an early age.

My cousin Andy's input about fencing really stuck (no pun intended), and I saw it as a way to incorporate the old with the modern. In this period in history, we did not have cellphones; computers were huge and inaccessible to regular people. There was no internet. So, how did we conduct research back in the day? We had phone books with the yellow pages you could use to look up businesses. You could call information on your rotary dial phone (push button, if you were lucky). We would go to the library, check out books to read, and take notes. That is exactly what I did. I went to the library, and read every book I could find about fencing. I went through the yellow pages, which was frustrating because every time I looked up fencing it referred to chain link, privacy, picket, or stockade fencing. Which brings me to a funny story.

When I was in my thirties, my parents put in a pool, and I helped with the landscaping. They ordered some concrete for around the pool. When the concrete guys arrived, I was installing a vinyl stockade fence. Unknown to me, one of the guys at the concrete place was an old fencing teammate from Notre Dame, and he recognized my name. He passed the word to his crew. The crew asked my name and I told them. They asked if I was the same Todd Griffee who fenced at Notre Dame. When I affirmed it, they had a field day with, "Look at the fencer building a fence!"

As I said earlier, having a sense of humor is important. It is important to make your journey fun and lighthearted. The quest for your goal should not feel like drudgery or punishment. It is supposed to be hard work, but rewarding.

Back to the research. I was having difficulty trying to find a place to receive fencing lessons or even order fencing equipment. It would have been a normal response to tell myself there were no fencing clubs, or places to buy the equipment, and move on to something more accessible. My heart and gut would not allow me to do so. The conviction I had of being on the right path was unwavering. I had a notepad, and would call places to get what information I could. After weeks of research, I located some businesses that sold fencing equipment. I had to call them and give them my address to send a catalogue (Remember, we did not have the internet).

Once I received the catalogue, there was another obstacle: How would I afford to order anything? I worked my tail off riding my bicycle around town mowing lawns, weeding gardens, and doing any other odd job I could find. When I finally saved up enough money, I ordered my first fencing sword, and waited impatiently for weeks for it to arrive. I still vividly remember the day it came. It made this whole plan seem more real.

The next hurdle was how would I learn to actually fence? I was finally having some success in locating Notre Dame's fencing coach, Michael DeCicco. Surely, he would know where I could find fencing lessons. After leaving numerous, unanswered messages with his office, I was feeling a bit discouraged. This would have been another opportunity to give up and quit. However, my resolve to accomplish this goal kicked in. I called again, and Coach DeCicco answered!

The ability to troubleshoot is a valuable lesson—and once mastered—will provide a lifetime of success. There is always a solution to a problem; it just depends on what you are willing to sacrifice to attain your goal.

Activity

The activity for this chapter is to review your own life or project, and identify the stumbling blocks. Write them down, and start researching how others successfully negotiated each of those obstacles. Write down every possible way to solve the problem. Good or bad, write the ideas down. Do not be afraid to pray to God for guidance and listen for the answers while being still and quiet.

I used to keep a notepad next to my bed because I would get ideas when I was dreaming. I would quickly wake up and write them down so I would not forget. I actually executed that same activity in the writing of this book. Be creative and resilient in discovering the solutions for the obstacles in your path.

_____

_____

_____

_____

_____

_____

_____

_____

_____

_____

_____

_____

_____

_____

_____

_____

_____

7

EXECUTE THE PLAN

My wife, Bonnie Jo Griffee, in 2018,
at the Grotto at Notre Dame on game day.

It is one thing to organize some ideas on paper, and a very different animal to breathe life into them. It can be intimidating, because it is no longer a story written on paper, it is becoming something real. This is where the excitement must overpower the fear, so you bravely push forward.

I left off in the previous chapter where Coach DeCicco answered the phone. This coarse, raspy, loud Italian voice boomed asking what I wanted. I explained it was my dream to learn how to fence. It seemed like an eternity waiting for his response. He replied that the team had just started conditioning, and ordered me to show up at the fencing gym at 4:00 the next day to see if I had what it takes to make it. That was the whole conversation. *Weeks of getting run around in circles, and tomorrow I was to show up!* I was thrilled!

At 13 years old, I was tall and lanky for my age. I was athletic from playing soccer as a goal keeper, riding my 10-speed bike, swimming, archery, and boxing. I showed up for practice the next day, and told the people who looked like they were in charge that Coach sent me. They told me to fall in at the back and we started calisthenics and stretches. Then we did footwork. I had no clue what footwork was, and others helped me along. Moving back and forth with your feet at 90-degree angles was bizarre, and was extremely strenuous.

Words cannot describe the amount of leg strength it takes to advance, retreat, and lunge in fencing. One's legs are not built to stretch that way, so you have to mold the muscles to do so. Then we would run four miles around campus. If you have been to Notre Dame, you know how beautiful and special this campus is. It was early fall, and I still recall the mix of smells from the turning leaves and fresh-mowed grass. The campus was teeming with students playing football in the Quads.

The runs would take us past the Grotto, which is a scaled replica of the grotto, where Our Lady of Lourdes appeared (Sanctuaire Notre-Dame de Lourdes, France, 1858). We continued around St. Mary's and St, Joseph's lakes, then back around to the Athletic and Convocation Center (ACC). Every time I smell campus during this time of year, it takes me back to that experience, and is a reminder just how special Notre Dame is.

The workouts were no joke, and I was holding my own. This was my opportunity to show what I had. Everyone was helpful, and the armorer (the individual who builds and repairs the fencing weapons), Roland, was probably sick of all of my questions. However, he patiently answered them.

After a month of training, someone figured out I was not a student there; I was in middle school. I recall feeling pensive about what was going to happen to me. *Would they kick me out? Would my dream be over?* The fear of losing this opportunity hit my stomach like a mule kick. I was told my parents and I were going to have to have a face-to-face interview with Coach DeCicco.

Coach explained he was impressed with my drive and hard work. My hard work paid off because he offered to put me in the novice program, which was a class designed to transition people from the physical education fencing class to determine if they could make the cut to join the varsity team. I gladly accepted the offer, and Marc became my instructor.

Marc was a fantastic instructor, and thinking back, he was under no obligation to help me. I am very grateful for his assistance. He took a lot of time giving me lessons and preparing me. After dabbling with all three weapons—epee, foil, sabre—it quickly became apparent the epee would be my weapon of choice. Marc also put me in touch with a group of guys who fenced there after hours as the Notre Dame Fencing Club. The guys in the Notre Dame Fencing Club were fun, and participating with them helped me train all summer long. When the school year ended, I met with Coach again and he told me I made the cut; I would be training with the varsity team the next season!

When life presents opportunities, you need to seize them and put in every ounce of energy. People get scared when opportunity knocks and they try to have one foot in and one foot out to cut their losses in case things do not work out. That is always a failure. As I mentioned earlier, change of any sort scares people; good or bad, it scares them. I know it sounds silly, but like most people, I was succeeding and I was scared I was going to screw it up. This is where I started visualizing my success.

I would imagine myself executing the proper moves. I would imagine myself making good hits, and winning bouts. I imagined myself winning tournaments. I visualized it so hard those experiences became super-realistic. In fact, they became reality. I would use this visualization of my success throughout my life. Writing this book has reminded me that every major achievement I made was preceded by extreme visualization.

Activity

The activity for this chapter is to review opportunities that you encountered or experienced. Write down how you reacted to them, and what improvements you could make to ensure your progress when the next opportunity comes to you. After you have written down these opportunities, visualize yourself succeeding! If negative thoughts of self-doubt enter your mind, quickly push them out.

For every negative thought, think three positive ones. Make this a daily practice!

_____

_____

_____

_____

_____

_____

_____

_____

_____

_____

_____

8

DO WHAT IT
TAKES TO MAKE IT HAPPEN

St. Mary's Lake, Notre Dame (courtesy of Todd Griffee)

When the excitement of a new beginning wears off, the mundane seems to slap you in the face. There is work to do; lots of work to do. As I mentioned before, faith without works is dead. You need to take action, and a lot of the time, it is work that does not have anything to do with your project. I was 14 years

66

old when I was given a golden opportunity, which translated to a lot of blood, sweat, and tears. It was not enough to go to practice, then go home, and expect to compete at a high level. There was always more to do.

After practice, I would work on more mundane footwork. Then I would run a four-mile course around campus, which included running a figure eight around St. Joseph Lake and St. Mary's Lake. I keep mentioning running around those lakes because it was such a magical experience. Running around those lakes and being in nature helped me connect with God. It was a very powerful experience.

It is amazing how therapeutic the beauty in nature can be. Despite the heat or the bitter coldness of winter, I ran. Autumn was particularly beautiful, and spring was welcome after the bitterness of winter. Winter was a Spartan experience to toughen me up for the challenges which were to come.

I would also target practice. I know you are wondering, *how can you target practice in fencing?* Most fencers put a mark on a padded surface, and hit it with the tip of their sword. The problem is, it does not move as an opponent would move. Rarely is an opponent going to just stand there and let you hit them. In the gym, they had golf balls hanging from a chain. The golf ball would be the same height as an opponent's arm, and you would hit it with the tip of the epee. Now it is bouncing all over the place, and you have to keep hitting it on the fly. At home, I had a golf ball hanging from a string in our living room. I got to where I could work that golf ball as a boxer would work a speed bag. I would also face a wall, throw a tennis ball at the ground in front of the wall so that it would bounce off the wall straight at me, then I would hit it with my point. This exercise really helped my distance and timing.

Have you noticed that all of these extra efforts are basics? Running for physical fitness, legwork for mobility, and target practice for accuracy—they all built muscle memory you could rely on when you were so exhausted you could barely hold up your weapon. They also were an investment. The more you invested in your training, the harder it was to give up or lose. When I lost, I would feel physically ill because of all the time, money, and effort I'd invested.

Speaking of money, fencing equipment was not only hard to find, it also was expensive. We did not have the internet, and I was too young to drive. My parents could not afford the equipment, so it was up to me. As I said before, I rode my ten-speed bike all over town to mow yards, weed gardens, paint, and any other odd job I could find. I look back and am exhausted with all of the work I did during this part of my life.

I think weeding large gardens for peanuts of money was probably the worst job. I even worked as a dishwasher at St. Mary's Convent when I was old enough to work with a permit. On one occasion, I cleaned out an entire apartment that was used for storage (took me two full days) just to get a pair of German Adidas™ fencing shoes. As I got older, and was able to drive, I picked up many jobs, to include busing tables at the University Club of Notre Dame to help towards my education and training.

Education was also a huge responsibility. Notre Dame is extremely competitive, and it was imperative that I earned good grades. I had attended St. Mary's Catholic Elementary School, but it only went to 6th grade. Therefore, I attended a local public middle school. Some of the kids were really cool, but others could just be bullies. I was serious minded, and hung around with the smart kids. In my mind, I was there to get good grades so I could attend Notre Dame.

One day, one of these kids did not like something I did, so he challenged me to a fight after school. I was not about to have my honor questioned, so I agreed. During the next class, another kid challenged me to a fight. I told him as soon as I was done with the first one, he would be next. Then in the next class, a different kid challenged me to a fight so I advised him he would be next after the second fight. Now I had three fights booked for after school. Before I could make it to the first appointment, the 8th grade bully jumped me, and the fight was on. I fought like the third monkey running up the ramp to Noah's Ark and, brother, it was raining hard. As soon as I settled that matter, I showed up to my first appointment, and my opponent thought it was better if we became friends. I agreed. That happened with all three appointments, and I never had a problem with bullies again.

There were also some really good teachers at the school, and I enjoyed their classes. I remember sitting in music class, and the other kids were making fun of me because I loved classical music, and enjoyed learning all about it. The teacher overheard it, and intervened. He told those kids that they would not be making fun of me when I was making over $100 grand a year and they would end up pumping gas for a living. They stopped the heckling.

Next up was high school, and I attended the local public high school. Now it was getting real, because these grades would determine if I would get accepted into Notre Dame. During the first semester of my freshman year, the math teacher, who was also the cross-country/track coach, kept telling me to go out for cross-country. I politely explained I could not because of fencing; he would belittle me and my participation in fencing. I soon found myself pulling a D in his class. I was beside myself because I was not going to Notre Dame with a D. I asked him if there was extra work, I could do. He responded that I could go out for cross-country. He was marking points off my homework because I was not stapling my papers in the side he wanted.

Finally, my mother got involved, and we met with the principal. The principal explained that the math teacher was the head of the math department, and I would be required to take his classes three out of my four years. There was no alternate teacher or course. The principal also explained that because the math teacher was tenured, he could not do anything with the teacher. He reported there were many complaints filed about this teacher, but there was nothing he could do.

My mother asked how I was supposed to get to Notre Dame, and the principal advised us to go to another school. My mother immediately withdrew me from school, and we drove to Coach DeCicco's office. We explained the situation to Coach, and he agreed I would never make it into Notre Dame with a D.

He called St. Joseph High School in South Bend, and made an appointment for us. He stated a lot of Notre Dame faculty and staff send their children there. Coach wrote a letter of recommendation for me, and after lunch, we met with the Registrar at St. Joseph High School. I handed him Coach's letter, and was immediately accepted. The Registrar filed the paperwork for financial assistance from our local parish, and I started school there Monday.

St. Joseph was like a breath of fresh air. I loved everything about the school, and knew I was on the right track. After the local parish paid a portion of my tuition, my parents made an agreement with me. I was required to pay half of my tuition, which I gladly accepted. I always felt good about that arrangement because they could not afford the tuition, and I felt more invested in my education.

The whole lesson of this chapter is to do the basic and mundane things that will be the foundation for your project. Later in life, I opened a British pub. It required me to do a business plan, plan marketing, draw plans for the building and health departments, and design a restaurant to comply with the Alcoholic Beverage Commission. I had to be a carpenter to remodel the space, an interior designer to decorate the pub, a janitor to clean it, a cook to design the menu, a bartender to properly pour a Guinness™, a server to market our products, and a supervisor to coordinate all of the employees. None of those are the glamorous tasks associated with owning your own business. They are the mundane, behind the scenes requirements for a successful foundation.

I remember chatting with a very talented professional musician about his career, and his plan for making it big. I noticed his plan lacked the mundane, behind-the-scenes requirements. He was not willing to do anything other than practice his music. I was immediately reminded I learned all of those lessons while preparing for my fencing career at Notre Dame.

Activity

Your activity for this chapter is to make a list of all the basic and behind-the-scenes requirements that are essential to your success. Once you make the list, start writing down solutions to obstacles standing in your way. The key to success is being able to do the mundane, unenjoyable, and basic things required to make your dream happen.

9

MENTORS AND SUPPORTERS

Mike DeCicco coached Notre Dame's fencing team from 1962-95.

Coach DeCicco (courtesy of Notre Dame Fencing Program)

The plan was moving along well. I was able to work my tail off for the necessary equipment. I was training with the varsity team, and the next step was to prepare for competing in tournaments. It is one thing to spar in practice, and a totally different animal to compete in a tournament. Tournaments could last an entire day, depending on how well you did. They would generally start early in the morning, with each competitor placed in a pool of six fencers. Within each pool, each competitor fenced the other five competitors. The fencers with the best win/loss records would advance to another pool of six, and so on, until you reached the double elimination bracket. The bouts in the pools of six would go to five points or touchés.

In the elimination bracket, your previous win/loss records seeded you, and the bouts went to 10 touchés. You usually had to be defeated twice to be eliminated. The bottom line is, there was a lot of fencing in a single day. The uniform and mask made you just drip with sweat, which made for an even more exhausting day. The trick was to keep your muscles warm, without overheating, and to drink tons of Gatorade™. Cramping was common, so we ate a ton of bananas.

Coach DeCicco spent a lot of time giving me lessons. These were not just technical lessons; they were physically brutal lessons, which lasted 20-30 minutes, and left me depleted. I literally thought my legs would collapse, and my arm would fall off. How could this 50-something year-old-man run a 15-year-old kid into the ground? Coach was an absolute dynamo, an Italian beast who never tired, and yelled the entire time.

However, those lessons were incredible. His ethos and persona clicked with mine, and he was a great mentor. He was really able to bring out the best in me. He took mentoring me to a new level. He became my sponsor for Confirmation into the Catholic Church. I even took his namesake, St, Michael the Archangel, as my Confirmation name. He loved reminding me of this, especially when he was dressing me down when I was called on the carpet. I was truly blessed to have Coach as a mentor.

When I was 15 years old, Coach felt I was ready to start getting some competition experience. Notre Dame hosted a qualification tournament for Junior Olympics. Believe it or not, there was actually some good competition. Culver Military Academy was not far, and Notre Dame and Tri-State University Students under 19 competed. This was my first outing after two years of preparation. I still remember the jitters during my first bout. It was like stage fright, but was compounded with the stress of being in a fight.

Fencing is similar to boxing and martial arts because you are literally fighting your opponent. It is a challenge keeping your emotions in check to stay focused, and the jitters were not helping, but I managed to keep it together by focusing on Coach's lessons. Coach's methods had prepared me well. I got my first win! In fact, I won the under 16-age group, and almost qualified for the under 19-age group. This meant I qualified for Junior Olympic National Championships in Van Nuys, CA. I was going to Nationals!

As with every other achievement, this came with more hurdles to overcome. How was I going to afford to get to Nationals? My father just started a new sales job, selling x-ray equipment to hospitals, and he worked on weekends at Jay's Lounge. My mother worked full time as a server at the University Club. They could not afford my trip, and all of my money went to the necessary equipment.

The first thing I did was ask Coach what I should do. Coach was able to get Notre Dame's discount on plane tickets and hotel reservations so that brought down the total cost of my mother and me going to California to approximately $1200. That was an incredible help, but now where to come up with $1200? Back then, $1200 dollars was a lot of money!

Like I said before, my mother was my biggest cheerleader, and she was a determined woman. If anyone could figure this out, it was she. Mom had a bartender friend named Lana, who worked part-time at a bowling alley in South Bend, IN. These two cooked up a plan. They had raffle tickets made up, and the bowling alley was kind enough to let us sell them. Those two sold, and sold, and sold raffle tickets. On a couple of weekend evenings I helped them sell raffle tickets. I was amazed with all the questions I received about fencing. Luckily, I had taken my fencing equipment because people wanted to see it, hold it, and act as if they were swashbuckling in the movies! However, the people were genuinely interested and they purchased lots and lots of raffle tickets. They all wished me luck, and it was very encouraging. We had the big drawing, and we had our $1200 when it was all said and done. I am so grateful for everyone who made that trip happen because it was life changing.

Leaving frozen Indiana in February to go to Los Angeles, CA was magical. The sun was vibrant and warm. No perm-a-cloud. The smell of fresh soil and cut grass was everywhere. The warm air and palm trees were a wonderful break from the bleak snow back home. We stayed at the Sheraton Universal™, which was exciting because we saw all sorts of TV personalities. Every morning, the hotel provided fresh-squeezed orange juice that was delicious. It definitely was not the powdered mix which the astronauts drank! We were invited to all of the team dinners, and a Notre Dame Alumni donated tickets to Disneyland™ and Universal Studios™ to all of us. This trip was a fairy tale.

When my competition day arrived, I was so excited. These kids were the best in the country. I recall wondering how I would compare. Well, I did more than compare; I finished fifth, and received a trophy, which still sits in my office, and a bottle of non-alcoholic sparkling cider. I was confident that I found a great starting point to improve upon. That trip changed my life because I worked hard for two years, and was fifth in the country in my age group. I had something to show for my hard work. It made me hungry to work harder to become even more successful.

I am most grateful for my mentors and supporters who made that trip possible. That trip was the first stepping-stone into my fencing career. Oftentimes, people do not have family, friends, or a positive support network. Not all of your supporters and mentors are going to just fall into your lap, and you cannot use that as an excuse. If you do not have family, friends, supporters, or mentors, you need to find them. Seek them out because they are treasures in your life. Remember in the previous chapters, Coach DeCicco did not seek me out. He did not even know I existed. I had to put in a lot of effort to seek him out, and prove myself worthy of his training.

Activity

Your assignment for this chapter is to make a list of all of your natural family, friends, and supporters. Let them know how much they mean to you, and thank them. Next, make a list of mentors, who would desire to have in your life. Then, go out and try to build a relationship with them.

_____

_____

_____

_____

_____

_____

_____

_____

10

CONFIDENCE AND PROFICIENCY

Competing nationally

Todd Gritten, of Niles, holds his epee sword that helped him into the United States Fencing Association Junior Olympics. (Daily Star photo by Roger Schneider)

5th Place (courtesy of The Niles Daily Star)

That success I acquired in California really fueled my fire to excel. I trained even harder and competed in every venue I could find and afford. I recall Coach telling me the only way I was going to improve was to compete against fencers who were

81

much better than I was. That was excellent advice, and I've used that philosophy throughout my life. By this time, I was 16 years old, Notre Dame had brought in some ringer epee fencers - Ola and Jan from Sweden. They were fierce competitors.

Ola was six-foot-five and could lunge from one side of the strip to the other. Once he got ahold of your blade, there was nothing you could do, because he was so ungodly strong. I made it a point to bout with him as much as possible. At first, it was an impossible task, but then I worked up to where I could score once or twice on him before he annihilated me. Eventually, I got to the point where I could win a bout; not every time, but maybe one out of five times. That was a big deal because this guy was invincible. He went on win the NCAA Individual National Champion that year. I learned a lot from him and gained a lot of much required confidence in the process.

A positive attitude is essential. You have to be confident that you can beat any opponent. Confident, but not arrogant. The minute you underestimate your opponent, you have set yourself up for a big failure. Competing against people much better than you is a healthy way to stay balanced between confidence, and arrogance. It also means that you cannot be intimidated. Once again, this is where visualization comes in handy. I would imagine competing against these gifted athletes and winning. I just kept believing until it became reality.

During that year, my father took me to a big competition in Cleveland, OH. The tournament was hosted at a very swanky hotel, and there were some top shelf competitors there. I hooked up for my first bout, and the director said, "Allez" ("go" in French). I ran full speed at my opponent, flèched, and buried the tip of my epee into my opponent's chest. I was up one to nil, and feeling pumped. My opponent hit me in the hand five times to win, without breaking a sweat. I remember being agitated, thinking, *what the hell is wrong with me?*

My father asked why I was upset because he had this great big smile on his face. I was thinking to myself, *why is he smiling? I just got destroyed.* He said the reason he was smiling, was because I scored on the World Champion. He had been talking to him previously, and he was proud of me because I was not intimidated and scored on him as if he was any other competitor. I had no clue who he was, so it was a consolation later on when it sank in that I had the opportunity to score on a World Champion.

Confidence and a winning attitude are imperative to success. Opponents sense doubt and uncertainty like dogs sense fear or sharks sense blood. I've carried this lesson for the rest of my life, especially in law enforcement. A criminal can sense incompetence and fear like nobody's business. If you show them weakness, they will exploit you, and will not hesitate to send you home in a casket. If you show up with polished boots, a sharp uniform, and a professional self-assured attitude, they are less likely to push their luck.

Arrogance is when one thinks they are better than others are, and they enjoy putting people down to make themselves look better. Arrogant people are often ungrateful for the opportunities they have received because they feel like they are owed what they have. They walk around with an air of entitlement. It is not a pretty sight, and is contrary to codes of chivalry.

They also display poor sportsmanship. Arrogant people are full of pride. I am not talking about "pride" in the manner my father would tell me to take pride in myself, take pride in my work, or take pride in my family name. No, the "pride" I am talking about is the narcissistic pride in which you cannot admit fault because you are full of being selfish, self-centered, and self-seeking; it is the pride in which one thinks they can rely on themselves, and turn away from God for His loving help.

When I was young, people often thought I was arrogant because I learned how to be self-assured and confident as I became more and more proficient. I never thought I was better than anyone was, but perception was everything. I discovered humility was the difference between being arrogant and being self-confident. I had to learn to sincerely express my concern for others, and be quick to compliment them. I also had to be quick to admit my mistakes, take responsibility for my actions, and admit to myself, I did not have all the answers. I had to come to the realization that I still had a lot to learn. It was also a major lesson to be more outspoken with my gratefulness when others helped me.

Activity

Your activity for this chapter is to write down the masters you can learn from, and write down ideas on how to either compete or learn from them. Even playing cello, I listen to recordings of my favorite cellists, and try to play along with them. Then I will record myself, analyze my playing, and practice more. My favorite cellist is Sheku Kanneh-Mason from London, England. He plays with amazing emotion and is effortless in his playing. My wife took me to one of his concerts in Chicago, IL, and it was amazing. I was able to meet and chat with him. I thanked him for his inspiration to my own playing. That is the sort of networking and mentorship you need to pursue. You might need to be creative, depending on what type of career, sport, or activity you are attempting to improve. Next, write down ideas of how you can boost your confidence.

_____

_____

_____

_____

_____

_____

_____

11

LESSONS IN SPORTSMANSHIP AND LIFE LESSONS

Cavalier (courtesy of Fighting Irish Media)

I recall having a meeting with Coach when I was about 17 years old. I was so worried about my future. *How would I get into Notre Dame? How would I afford it? How could I improve my fencing?* I was training furiously, and it seemed like I was stuck on a plateau.

Coach, with his infinite wisdom, told me a story about crossing a stream. He said if you focus too much on the other side, you will slip off the slick stones you are using to cross the stream. If you only focus on the slick stone you are on, you will never see how to cross the stream. The key is to keep an eye on the stone you are on, look at the next stone, make the step, and look up occasionally to make sure you are heading in the right direction. He told me to focus on the daily activities and basics, and then look at what I could do in the first half of the season, and on and on. He reassured me everything would fall into place, and I would get across the stream.

He was correct. I did get across the stream. I have used this advice throughout my life. Whether I was trying to start my career in law enforcement, building my business, or re-entering law enforcement, this analogy always came back to me. It provided the necessary focus required to complete the projects I had undertaken—such a perfect analogy of an incredible life lesson.

He also discussed my temper. I had a fiery temper, and hated to lose. I was not John McEnroe (the tennis player known for nasty temper tantrums and throwing his racket), but I was bad. I was rarely mad at my opponent. I was enraged because I was not executing my plan properly, or did not have the required level of concentration. Coach also had a fiery Italian temper, so he understood me. He gave me pointers on how to channel that fire to win. Losing was miserable, but it drove me to work harder. When you are in the middle of a bout, and things are not going well, it is important to think about all of the work you invested to win. Thinking back, it would make me sick thinking about losing after I invested so much effort. It felt like such a waste. Coach helped teach me that if I lost, it was just as important to analyze and learn from those mistakes as part of improving, as it was to analyze and learn from winning. He really helped keep me balanced.

As a side note to that meeting, Coach gave me another piece of advice. He told me to walk around with an epee in my hand and hit every little cigarette butt, piece of trash, or stone I picked to target practice. Therefore, I kept an epee with a practice blade in my locker at high school, and when I walked to the gym at Notre Dame, I target practiced the whole way. The funny thing is, the school was not alarmed I kept a sword in my locker, but the people driving by probably had no clue what I was doing.

There is more to being a good sport than just shaking hands after you have been defeated or giving a sincere thank you to an opponent you just defeated. As I progressed during my junior and senior years of high school, I became proficient and could defeat anyone on the varsity team on any given day. It was getting to the point where I was doing a lot of winning. On one particular day, I was fencing a very talented fencer. I was toying with him, and being a bit tricky, and showy.

Even though I won the bout, Coach immediately called me into his office. He asked what the hell I was doing. I said I won. He lectured me on why I missed the point (no pun intended). He referred back to my bouts with Ola. Coach said, "Ola went after you like nobody's business. He gave you his best game so when you did score or win, it was a real measure of how you were progressing." Coach explained how it was poor sportsmanship to have toyed with that talented fencer. He ordered me to go out and try to defeat everyone in the gym 5 to nil. Give them my best game, just as Ola gave me his. That talk really sank in. It was such a simple lesson I had overlooked. I never made that mistake again.

The other thing Coach taught me was humility. Not the degrading negative context the word humility often has; this was the humility of being teachable. Coach DeCicco taught me to take harsh constructive criticism without becoming defensive and always trying to justify myself. This was admitting I made a mistake or needed to learn a better way. This is also a life lesson I have carried on to this day. They say you cannot teach an old dog a new trick, however, I pride myself on now being an old dog who enjoys learning new tricks. This was evident when I re-entered law enforcement. A lot had changed in law enforcement, from weapons to tactics, laws, investigative techniques, and technology. My being able to open myself up to being vulnerable to constructive criticism, and being humble enough to learn new techniques has been a great strength to my career, and I owe that lesson to Coach DeCicco.

Another thing I learned from Coach DeCicco was how to challenge and lift my teammates. Even though Coach could be harsh with his constructive criticism, he was always fair and did not put down team members. He had a talent for finding the best qualities in each fencer and always made a point to lift them up. This is the exact opposite of the arrogant person, who puts others down to make themselves look better. I did my best to apply those skills to the members of my squad when I was captain, and tried to carry out those principles in every leadership position I held afterwards.

The final lesson Coach DeCicco taught was to not compare myself with others. It was more important to measure my personal growth. The benefits were twofold: On one hand, I was not competing against my opponent directly; I was trying to improve my skillset from the previous day, which held me accountable for something I could control. On the second hand, it kept me from getting my feathers ruffled by an opponent whom I did not like. This concept helped develop personal responsibility and kept me from blaming others for losing.

Activity

The activity for this chapter is to write down nuggets of truth and life lessons you have learned from mentors, family members, friends, and support people. Write down how you think your journey is measuring up to those life lessons. It is also important to write them down so you never forget them, because history is cyclic. These tools will help you negotiate obstacles throughout your life.

_____

_____

_____

_____

_____

_____

_____

_____

_____

_____

_____

_____

_____

_____

_____

_____

_____

_____

_____

_____

_____

_____

12

DECISIONS AND SACRIFICES

The Defensive End (courtesy of Leslie Griffee)

My high school years were not completely Spartan with only training, fencing, and studying. They were fun filled years spent with good friends doing what 80s kids did. We cruised the strip on Friday and Saturday nights, went to dances, movies, out for ice cream, and went to each other's houses to play board games. The 80s was a great time to be a teenager. The fashion was cool, the music was great, and TV shows like *Miami Vice*© were fun. St. Joseph High School was a great fit for me, and I really enjoyed the teachers, classes, and students.

They had a very successful football program, and I loved football! I had been a huge Notre Dame Football fan since I was a wee one. The bug bit me, and in my sophomore year I went out for football. I became a starting defensive tackle on the junior varsity. The season was in the early fall so I only missed a couple of conditioning weeks for fencing, and I started the fencing season in tip top shape. Football was a blast, and I met many great guys playing. I conditioned extra hard the summer between my sophomore and junior years, and earned my spot as starting varsity defensive end. During two-a-day practices, I was running with the running backs in the 40-yard dash, and could hang with the wide receivers for the 100-yard dash. It was also super cool that our practice field was across the street from Notre Dame, and we could see the Golden Dome as we practiced.

Even though I was only six foot and 180 pounds, I was fast and aggressive. I loved playing that position. My coaches told my parents I was like Jekyll and Hyde. They said I was the nicest kid in the world, until I put on my helmet and was mean as hell. We had a successful season, and during our second playoff game,

I pulled some ligaments in my knee and was out for the remainder of the playoffs.

Luckily, it was minor, did not require surgery, and I was back to training in no time. The bad thing was that Coach DeCicco's intelligence network somehow found out I was injured in a playoff game. That earned me a trip to the hot seat in his office. He was livid that I was injured playing football. He made it abundantly clear that football had nothing to do with my goals of fencing success while attending Notre Dame. He gave me an ultimatum: Quit football my senior year, or do not go to Notre Dame. He also stated I was not allowed to box or play soccer. I was going to Notre Dame to fence, and was prohibited from any other extracurricular activities. He explained he had invested too much time in training me for me to get injured doing something silly. It probably did not help that one of his star varsity fencers injured his knee while skiing.

I knew Coach was right. It was a no brainer decision. Of course, I picked Notre Dame over football. During my senior year, the St. Joseph High School football team played a playoff game in the Hoosier Dome in Indianapolis. It was pretty rough watching someone else playing my position in this forum. That stung, but I used that sting to help me focus on my original plan to find success at Notre Dame.

There were also sacrifices of time because I had to practice or work and missed opportunities to hang out with my friends. Of course, there was also the sacrifice of having to set money aside to pay my tuition and pay for my fencing equipment.

There are times in your life when it is easy to get distracted from what your goal is. Life finds a way to put everything from luxuries to obligations in the way of the goal. No matter how much you try to balance everything, there comes a point when you have to decide to make sacrifices to complete your goal. It ties right in with the whole concept of removing the excuses. Sometimes you need to sacrifice the distractors. Those sacrifices are investments into your dedication to your goal. The more you dedicate, the harder it is to fail. As Ringo Starr said," It don't come easy." I was lucky enough to have a "Come to Jesus" moment, and get refocused.

Activity

Your activity for this chapter is to write down everything that interferes every day with working toward your goal. Those obstacles might be family, friends, responsibilities, money, time, luxuries, etc. Next, prioritize those obstacles, because some of them are important responsibilities which cannot be sacrificed; others are more like "wants" than "needs." Finally, look at what you can sacrifice so you can invest more in your goal.

_____

_____

_____

13

HOW DO WE PAY FOR THIS?

ROTC picture (courtesy of Notre Dame Army ROTC)

Senior year was exciting because I could see the plan coming together. Coach had advised me to look out for my acceptance letter into Notre Dame. I was accepted! To say I was excited is an understatement. I remember telling my favorite teacher, Barb, who taught English. She asked how I could know I'd been accepted; the letters had not gone out yet. When I showed her the letter, she was perplexed, but excited for me. My family was super excited as well. Contrary to Han Solo™ telling C3PO™ not to tell him the odds, I rather enjoyed the thrill of having the odds stacked against me. You know how many very talented people apply to Notre Dame, and how few are selected. The odds are definitely not good. For me this was a sign of God's grace. This was indeed a miracle, and one for which I am eternally grateful.

After the excitement settled, there came the realization of how I was I going to pay for this. Like so many times previously on this journey, with each opportunity came a new price tag. Notre Dame was ungodly expensive, and even though fencing scholarships were available for female athletes, they were not available for male athletes at this time.

During the previous year, I was following Coach's "crossing the stream" advice and looking ahead at this potential problem. I remember talking to the foil captain, Chris, who was in Navy ROTC. My plan from childhood always included serving in the military, but I did not see how it would fit together. Chris gave me some good insight, so when I received my acceptance letter, I made an appointment with the Navy commandant at Notre Dame to see if I could get an ROTC Scholarship. I made sure I wore a suit and was well prepared for the interview. It all made sense now. This scholarship would allow me to afford Notre Dame and provide me with military service, which could later prepare me for a career as a federal agent. This was the answer.

I arrived early, with a smile on my face, and was ready to crush this interview. I introduced myself to the commandant, and looked him in the eye as I firmly shook his hand. I was off to a great start. He had me tell him about myself, and I explained my journey with fencing and my acceptance into Notre Dame. He was less than impressed and commented that it was difficult for varsity athletes in ROTC. I regrouped and kept a smile on my face and a friendly tone in my voice. He then asked me what I wanted to do in the Navy, and I answered Special Warfare. He asked if I was an Olympic caliber swimmer. I answered I was a strong swimmer, and an Olympic caliber fencer. He replied fencing is not swimming. He asked what else I would like to do in the Navy, and I responded I loved flying and could be a fighter pilot. He asked what my major was going to be, and I responded that the Secret Service had recommended sociology. The commandant replied that if the major was not aeronautical engineering, then fighter pilot was out of the question. He made it clear he did not think the Navy was a good match for me and dismissed me as if I was a nuisance. I left furious.

The Navy had been my first pick because of my father's and uncle's positive experiences in the Navy. The irony was, I later went to work for the Navy in various law enforcement capacities, and the Navy, by far, has treated me better than any other employer. It just goes to show you cannot judge an organization by one member. However, in this case, the decision maker had spoken and the Navy was not an option. The plan was down the tubes and I had to go from Plan B to Plan C.

It would have been very easy to throw up my hands and give up right then. I came all this way to be accepted to Notre Dame, just to be dismissed like a pest. However, I checked my attitude, and would not allow myself to be defined by some uninterested party. I marched my ass right down to the Army ROTC Office, and asked if the colonel was in. They took me in immediately to speak with the colonel.

I explained what happened with the Navy and he asked how I was accepted into Notre Dame. I told him I was recruited through the fencing team. The colonel said the Army loves varsity athletes, and they would work around my schedule. He asked what I wanted to do in the Army, I said Special Forces, and he said, "Well you are definitely in good enough shape being an athlete." He had me take some tests, and stated I scored high enough to qualify for a scholarship, but I missed the deadline. I was supposed to have applied before I was accepted into Notre Dame, which made no sense to me.

The colonel had Coach DeCicco sit in on my scholarship interview panel, and I received a three-year full scholarship. They would also provide book money every semester, and a monthly stipend. The athletic department provided my books, tutoring, and evening meal. To get me through my freshman year, the Army hired me as a student worker to work 20 hours a week, and that money was applied to my tuition. A donation from Dad's beer wholesaler and student loans were able to get me through the first year.

By this time in my life, I was figuring out God was a show-off. He liked to wait for it to be a fourth down passing situation, and then throw for a touchdown. He certainly figured out this dilemma for me. Being accepted into Notre Dame was a miracle in itself, but paying for it—now that was a miracle. It seems life likes to throw adversity into the mix to get us to give up. I think God does it just to see how dedicated to the goal we are, and how reliant we are on Him.

A priest once told me that if you have a vision of something you want to do, and it is easy to complete, God is not necessarily involved. However, if your goal is impossible, and you are at a loss how to complete it, then it probably comes from God. It is definitely a miracle when it happens.

This is one of the reasons I wrote this book. It is not to show how great Todd Griffee is; it is to show how God worked many miracles in Todd Griffee's life, and if He can work miracles in my life, He can work miracles in your life. This is true hope.

Activity

Your assignment for this chapter is to pause on your journey and look ahead to see where you are going. Write down what obstacles you expect to face crossing the stream. After you have your list, write down possible ways to negotiate those obstacles when you get to them. When I was a police officer, and I was responding to a dangerous call, I would envision everything that could go wrong so I was mentally prepared when I arrived. Working for the Navy, we had such policies, standard operating procedures (SOPS), and pre-planned responses (PPRS) for about any incident imaginable.

This exercise is no different. You need to try to imagine anything that could go wrong and try to find an appropriate solution.

_____

_____

_____

_____

_____

_____

_____

_____

_____

_____

_____

_____

_____

_____

_____

_____

103

14

FRESHMAN YEAR

The Ring (courtesy of Todd Griffee)

Fall at Notre Dame is an exciting time of year; the beauty of the campus had all of the excitement and anticipation in the air. I was finally here as a real student. I looked back at all those extra miles I spent running around this campus, preparing for this moment. It was surreal, intimidating, and exciting, all at the same time. The insanely busy school year started with "Hell Week" for the Army ROTC Ranger Company. It was a week of arriving at zero dark thirty for inspections, physical fitness, and ridiculously hard activities, which were designed to improve discipline, motivation, confidence, and teamwork. I was certainly glad when I completed that week, and earned the honor of wearing that black beret with my ROTC uniform. I was obligated to take a military science class every semester, and I found them quite interesting. The Army followed their word and adjusted my physical fitness schedule during fencing season. The Army also worked around my schedule while I worked my mandatory 20 hours as a student worker. I was assigned to a sergeant first class, who really took me under his wing to get me up to speed faster than the other cadets.

Coming from a college prep high school, I figured Notre Dame would be difficult, but I would be well prepared. Notre Dame was much more difficult than I had anticipated. It was like shell shock trying to get into a rhythm with these classes. Finite math was ridiculously difficult, but the professor tutored me and got me through it. The calculus professor was from Poland and I could not understand him. Once again, I needed tutoring. Because I took French in high school, I was placed in a higher-level French class. The professor thought it would be cool to deviate from the university grading scale. If you got an 80%, it was a "D." To add insult to injury, if a sentence was worth 2 points, and you made 4 mistakes, you got a minus 4. You could actually miss more than 100%. That was my only "D" at Notre Dame, and earned me a spot in entry level German. I thought earth science would be a breeze with my interest in paleontology. Wrong! They were teaching old theories that dinosaurs were cold blooded reptiles, the professor gave us reading assignments and lectures that contradicted themselves, which meant no matter what you answered on the exams, you were wrong. Thank God I had his student assistant as my tutor or I would never have passed. That was the thing about Notre Dame. Athletes really had to study and do their own work. Coach DeCicco was the academic advisor for all athletes, and when your grades dropped below a "B", he was informed, and you were assigned a tutor. It was not uncommon to have a Heisman Trophy Winner, and many other football players who would end up playing with the NFL on Sundays, in your study group. The tutors did not do the work for the athletes; it was more work for the athlete. I have to give Notre Dame's integrity credit because athletes were not given a pass on their grades. They actually had to work harder to earn them.

Fencing was the longest season of any varsity sport. We started conditioning after the first month or so of fall semester, and it ended with Nationals during mid-March of the following semester. Practice was at least two hours, and often went two and a half hours. I also spent extra time after practice running or doing extra drills. Then it was off to eat, and study into the late hours. The team travelled some before Christmas, but most of the travelling was January through March. On weekends when we did not travel, I worked as a busboy at the University Club with my mother. Once the fencing season was over, I resumed early morning runs and field exercises with the Army. To say my schedule was full was an understatement. I look back and I am amazed how I ever did it. The biggest payoff during that year was the team winning an NCAA National Championship! This included earning a very large ring.

Activity

Your activity for this chapter is to write down your busy schedule, and organize it. When you are feeling overwhelmed like there are not enough hours in the day, it is important to write out a schedule to keep you on track. I found it useful to make a checklist of everything I needed to accomplish that day. It also helped give me a sense of accomplishment, when you see progress completing such a heavy load.

_____

_____

_____

_____

_____

_____

_____

_____

_____

_____

15

BURN OUT

Fencing headshot (courtesy of Fighting Irish Media)

I spent the summer between freshman and sophomore year installing fire prevention and security systems in restaurants and factories for a local company. I also mowed yards, and worked several nights at the University Club. Summer was over in a flash, and sophomore year started with the same intensity as the previous year. Even though I did not have to work as a student for the Army anymore, I picked up more hours at the University Club. The classes were equally as challenging, and practice was exhausting. By the time I reached the end of fall semester, I was spent. I was burned out, and could not imagine continuing this madness. I quit.

I was fully prepared to walk away. My buddy's dad, Dr. Koellner, was a professor at Notre Dame, and he heard I was quitting. He contacted me and had a heart-to- heart chat with me. He explained that freshman and sophomore years are the two hardest years, and the last two years are much easier because you know what is expected, and you have completed all the required core classes. He also told me the importance of a Notre Dame education. Dr. Koellner stated it was something that no one could take away, and when life knocks you down, you can always look back at this major accomplishment. He said it would be the greatest regret of my life to drop out. He advised me not to make my decision yet, and to think it over during the Christmas break so I could make an educated decision after I had some much-needed rest.

I followed his instructions, and after the break I decided to stay in school. I figured I was almost half way there, so I could tough it out and hope it got easier. Dr. Koellner's advice was some of the best I ever received. He was absolutely right. My last two years were much easier, fun, and rewarding. It scares me when I look back at how close I came to giving up this dream. There have been many challenges since, when I was able to look at my Notre Dame education and use that as inspiration to negotiate other dilemmas in my life.

The other important factor was having a strong support system. My father said he would support any decision I made, and my mother, being my greatest cheerleader, was there to encouraging me. I also had wonderful, supportive grandparents and friends. I cannot express enough how important a strong support system is to completing difficult tasks, and dealing with the rigors life can throw at you. If you do not have a support system, please find one. There is no need to suffer alone through life's trials. God did not design us to be alone.

Activity

Your activity for this chapter is to write down mentors and support people to seek out when you hit your breaking point. When the chips are down, it is important to have the contact information for those you can rely on to give you good advice.

_____

_____

_____

_____

_____

_____

_____

16

BALANCE

En Guarde (courtesy of Fighting Irish Media)

As I said, after the much-needed rest during Christmas break, I returned refreshed and ready for the challenges ahead. During practice, a sports psychologist gave a talk to the team and stated he had methods for improving concentration and performance. I was intrigued and wanted a boost to my career so I made an appointment. The first thing I learned was meditation. His guided meditation sessions included imagery, where you envision executing your moves effectively. You would also envision yourself winning. This was powerful medicine. He recorded a session for me so that I could listen to it and meditate before practice and competitions. I was already familiar with using my imagination to visualize myself winning, but this was a totally different animal. My focus became laser-like, my concentration was powerful, and my confidence was unreal. This stuff really worked!

The psychologist also taught me about balance through focusing on improving my mind, body, and spirit. It finally dawned on me why my first two years were so rough. There was no balance. It was just work, work, work, and the only joy I felt was just keeping my head above water. I took measures to increase my spiritual relationship with God. When I ran around campus after practice, I made it a point to stop at the Grotto and pray for strength to continue. Even now, I try to get there to light a candle for my mother and grandparents, and I am reminded of those days. I remember wiping the snow off the kneelers to pray in the bitter cold, and then forge onward to complete my workout. I did not feel alone anymore. I also made sure I worked in some breaks to have some fun, and unwind. It was important to find time to decompress and let myself heal and rest. I also failed to do that the first three semesters.

On Sundays, I would watch some TV or a movie to relax my mind. I also started taking guitar lessons, and later switched to bass. My senior year, I actually played in a little church group during Mass. Music is a wonderful mechanism to relieve stress, work the mind, improve hand/eye coordination, and provide a spiritual outlet. I have carried this lesson to the current day with my cello playing. The benefits of learning a musical instrument are immeasurable.

This idea of balance, meditation, and working smarter really was paying off. I moved up to being a varsity starter, and was destroying my competition. Life was good! To think about how close I came to giving up, compared to the amount of success I was achieving at this point, was mind-boggling. How quickly events in your life can change when you change your attitude and perspective.

Activity

A dull axe does not do good work so it is important to take the time to sharpen the axe to get a better work product. Your activity for this chapter is to write down ideas for how you can improve your balance. Look at ways to improve spiritually, mentally, and physically. Also, make time for some fun and relaxation. There are many excellent guided meditation videos online that are extremely beneficial to performing at your highest level. I encourage you to meditate on a regular basis.

My first All-American Award, NCAA Championships
(courtesy of Fighting Irish Media)

THE WEAK LINKS

The Weak Link (courtesy of Leslie Griffee)

The euphoria I was feeling from my success quickly
ended when I read the Notre Dame student
newspaper, *The Observer*. There was an article about our
upcoming fencing season and Coach was quoted as
saying the epee squad was the "weak link" for the

upcoming season. That really stuck in my craw. The season had barely started and my squad was being called out as the weakest of the three. What a bunch of rubbish! The more I thought about it, the angrier I got, and then it dawned on me; this was a golden opportunity. I could exploit this as a real motivational tool to inspire my squad.

I had t-shirts made up with three links, a broken one in the middle. I gave one to each member of the epee squad. We showed up at practice sporting our new shirts, and weak link headbands. It became our mission to prove Coach wrong, and we wore our "weak link" shirts as badges of honor. The latest article in *The Observer* told the story of how the epee squad crushed our arch-nemesis, Wayne State University. This motivational tool resulted in my qualifying for NCAA championships, and defeating an opponent from Columbia University to earn All-American honors. The team elected me as captain for my junior year.

Looking back on this, I almost wonder if Coach did not orchestrate this from the beginning. He probably knew his comments would provoke me to excel. Either way, it worked!

During my life, I have heard many stories of how people have reacted when others put them down. Some people ignore it; others take the comments to heart and quit. However, the real winners latch onto those comments and make it a life mission to prove those comments wrong. I had this exact conversation with my partner, Jason, at work. He relayed the story of when somebody told him he would not be selected for chief petty officer his first time up. Jason made it a point to prove him wrong, and was selected CPO first time up. Winners find a way to defy the odds.

Winners have a way of viewing hardships as opportunities. When the chips are down, this is your opportunity to show what you have. I didn't learn the tools I used to turn the weak links into the strongest links overnight. They were byproducts of millions of baby steps taken over many years. It is my hope you can take the lessons from my story to become the strongest link in your journey!

Activity

Your activity for this chapter is to write down obstacles in your path, then write down how you can turn those obstacles into opportunities.

_____

_____

_____

_____

_____

_____

_____

_____

_____

18

DEFEND YOUR HONOR

The Duel (courtesy of Fighting Irish Media)

During summer break between sophomore and junior years, I installed fire prevention and security systems on week days and worked at the University Club nights and weekends. Despite the busy workload, I found time to relax with friends. One of my favorite memories was going to the dunes on Lake Michigan. Lake Michigan is gorgeous during the summer and I still make it a point to go there when I am in the area. After a busy, yet rewarding summer, I was anxious to meet the new challenges of my junior year.

I was taking some acting classes that were interesting. I remember telling a friend of mine that the two classes I use the most in my career as an NCIS investigator are English—because it requires a great deal of writing—and acting. He asked why acting, and I replied because when you are interrogating or interviewing someone, you are always acting. You have to act as if what they did was not so bad and you are their friend, or you have to act as if you are angry to invoke a response. Either way, it is all acting. Most of my classes were also honest classes. If you worked hard, you got a good grade. They were also more interesting, unlike many of the earlier required core classes had been. I remember having a professor for military history who was a fantastic lecturer. I could not wait to get back to class to find out what was going to happen next. He made us feel like we were right there in the mix of it all. I also had a wonderful English literature professor whose lectures I loved.

I was enjoying the leadership role as captain, and really liked the "weak link" crew. They were all hard workers and fun to be around. They were also very competitive, which helped elevate everyone's game. They also walked around with the same "weak link" chip on their shoulder, and bought into the concept that we were actually the "strongest links!"

We were at a tournament where we competed against many different teams. We were going up against a team which was not well known for being competitive in fencing, so I put a young, talented freshman in my place to face off with their captain. It was a freak of nature. The opposing captain dispatched my freshman 5 to 1. I was shocked, and the freshman was furious. On my way back from the restroom, I heard the captain of the opposing team telling a teammate all sorts of nasty things about Notre Dame, and bragging about his win.

You can trash talk me all you want; I have thick skin. The nasty things they said about "Our Lady" crossed the line. It was not just poor sportsmanship; it was insulting on a primal level and made me madder than a horned toad getting ready to squirt blood! When I walked past, I did not have my warm ups on so they had no clue who I was.

I switched the roster to put myself in against their captain as the last bout. Our assistant coach was livid because we had already won the match. This was not about winning. This was about defending the honour of "Our Lady!" It was one thing to brag about beating a teammate, and quite another to say these vile things about Notre Dame. This was about honor. This was a modern day duel.

Taking my position, I acted as if I did not even know how to hook up, or even how to fence. The split second the director said "Allez," I ran full blast, flèched, and buried the tip of my epee into his chest before he could even move. The second time, I ran at him and hit him in the hand, and repeated the same thing for the third touch. On the fourth touch, I ran at their captain, and he was frantically waving his arm back and forth so I hit him in the foot. For the final score, I ran at him and he was using his epee as a windmill, while he was tripping backwards. I cocked my arm back, and punched him in the face with the tip of my epee. The tip ripped a gouge across his mask, which left a bloody mesh imprint on his face. The tip was quickly followed by my bell guard, which knocked his helmet off, and sent him sprawling on the floor. His nose was bleeding, and I knew he was going to have a black eye. I quickly asked him if he was OK, and as I helped him up, I told him to have a safe trip home.

I know it does not sound very sporting, but there are lines one should not cross, and making rude comments about "Our Lady" was crossing that line. Some of you might be a bit squeamish and think this was a bit violent. In reality, fencing is a lot rougher sport than people imagine because you are, in fact, fighting so emotions can run high. It was not uncommon to get jostled (body slammed) by an opponent or get punched when we would come into close quarters. Some of the Swedish fencers were very physical, especially when they were losing. I once pushed a Swedish opponent over the director's table on the side of the strip after he punched me. I thought it was going to turn into a hockey match for a minute.

Trash talking in sports and in life are common, but it is not for me. There are too many times where I have seen it backfire. There is always someone better, faster, stronger, smarter. It is better to be quiet and give it your best shot. If you win or lose, you appear classy, as opposed to running your mouth and looking stupid after you lose.

Activity

Your activity for this chapter is to write down your triggers someone can use against you to provoke you. If you identify them first, it is easier to stay focused on the task. Also, write down things that are off limits. When someone crosses those boundaries, you can find great motivation to negotiate your task. Sometimes you need to make a stand for your core beliefs. Identify what you would fight for. Remember to stay humble in victory because there is always a bigger fish.

_____

_____

_____

_____

_____

_____

_____

_____

_____

_____

19

TAKE RESPONSIBILITY
FOR YOUR ACTIONS

Alone (courtesy of Leslie Griffee)

The wins kept adding up, but the assistant coaches became unbearable. No matter what I did, it was wrong; they micro-managed me to a fault. It was not a good blend with my fiery temper and hard charging demeanor. I had been very successful for the previous nine years, and was harder on myself than any outside person could be. The problem was that Coach DeCicco was not around as much. Tragically, his best friend's wife was dying of cancer, and he was busy consoling them. My trials were trivial compared to what was on his plate so I did not bother him with my troubles.

I dearly missed the support of Coach DeCicco; he was a fantastic mentor who really understood what made me tick. The relationship with the assistant coaches became stifling and toxic. It felt like they were just waiting for me to fail. Well, I did fail. At a tournament in Ohio, the men's team fenced on Saturday, and the women's team fenced on Sunday. The "weak links" had a great day on Saturday so on Sunday, we snuck out to a local pub to watch some football, and we got drunk. When we returned, the assistant coaches were furious. I knew I should have been there to cheer on the Women's Team, just as they cheered for us the day before. I knew I screwed up. The assistant coach told me he had told Coach DeCicco about the incident, and he was livid. I was told to stay home from practice Monday.

I think they really told me to stay away all week. That was ridiculous. My mother and father did not raise me to hide from my troubles. I made a mistake, and I was going to take responsibility for it, not run from it. It was nobody's fault but mine. On Tuesday, I showed up to practice, stood in front of the entire team, told them exactly what I did, apologized, and stated it would never happen again. After warm-ups, I was called into Coach's office and he read me the riot act. After he finished berating me, I agreed with him and apologized to him. I advised him I had already apologized to the team, was taking responsibility for my actions, and asked what else he wanted me to do. He told me I was staying home from the tournament in Madison, WI. I agreed, and stayed home. I did not think about it at the time, but those wins I missed would have probably placed me near the top of the most career wins in Notre Dame epee fencing history. That was a costly mistake, but as the character John Wick likes to say, "There are consequences." I deserved it and took my licks.

I redoubled my efforts and won the Great Lakes championship, and the Midwest regional championship, so I was double qualified for the NCAA national championship to be held in Princeton, NJ. The problem was that my relationship with the assistant coaches had deteriorated to a toxic level. I felt like I made a mistake, took responsibility, accepted my punishment, and kicked out two huge wins. Nevertheless, it seemed like they were waiting for me to fail. True or not, that is how I felt.

To add insult to injury, the team had a meeting the week before we left for nationals to elect team captains for the next year. In the nine years I had been training with this team, this always happened after NCAA national championships. This was odd indeed. To nobody's surprise, I was replaced as captain for my senior year. I had really loved this leadership duty, and I felt like I failed my position. I was feeling low, very low.

It was just a couple of days before we were leaving for Princeton, and one of the assistant coaches was giving me a lesson. My focus was terrible. I was distracted, and could not concentrate. I was desperately trying to pull myself together, but I could not have hit a barn if I was standing inside of it. I was angry with myself for not doing well. I was frustrated because I had the weight of nationals in a few days. The assistant coach stopped the lesson and told me if it were up to him, I would not be going to nationals because I did not deserve it. I am thinking to myself, *I just won two major tournaments to qualify, and I do not deserve it?* He then told me the only reason I was going was that Coach DeCicco overrode their decision to keep me back. The assistant coach kicked me out of practice and told me to go home. I was seeing red. I took my epee across my knee, snapped it in half, drop kicked my helmet across the gym, and walked out. The hardest part was Coach DeCicco would be unable to make the trip to Princeton, NJ, due to his friends' family emergency. I did not discuss the matter with Coach because he had more important things to worry about.

I discussed the importance of learning humility earlier, and I did my best to apply those skills in this situation. I opened myself to constructive criticism, admitted my shortcomings, attempted to make amends, took responsibility, and did my best to be teachable. In this situation, it was not enough. I was very hard on myself because I let myself down. I let my team down, and failed as a leader. I also had the pressure of competing successfully at NCAA national championships and despite the loss of my leadership position, I still had a duty to the team to do well. All of these outside stressors magnified my fear of failing, and it was important for me to save face by having a successful tournament. This lesson was beyond humility. It was eating humble pie.

There are times in life when your troubles become so great that you cannot see a way out. I remember, later in life, I was going through a crisis, and I told my friend, Mike, he promised God would not give me more than I could handle. I said I had more than I could handle. Mike told me God does not give you more than you can handle in a 24-hour period. He asked if I had a roof over my head, food in my tummy, and a way to get to work in the morning. I replied I had, and he stated, "Then you have what you need for this day." That was so true. Looking back at this dark week of my fencing career, I surrounded myself with family and friends to whom I was very grateful.

Activity

Your activity for this chapter is to make a gratuity list. When life dishes out more than you can take, make a list of all the things for which you are grateful. It quickly becomes apparent how much we do have in our lives and also keeps our focus on the positive and off the negative. A positive frame of mind is essential. For every negative thought you have, immediately think three positive things to keep yourself focused on the positive.

_____

_____

_____

_____

_____

_____

_____

_____

_____

_____

_____

_____

_____

_____

_____

_____

_____

_____

_____

_____

_____

_____

20

LEAVE IT ALL OUT ON THE STRIP

The Bout (courtesy of Fighting Irish Media)

The moment I landed in New Jersey, the assistant coaches started in with their restrictions. I was told after dinner that I was not allowed to leave my hotel room. This was insane. I had the weight of the world on my shoulders, and the next day was the biggest competition of my life. There was no way I could stare at the ceiling and let my mind race all night. I would be exhausted before the competition even started. Suddenly, there was a knock at the door, and a teammate said to come along; we are going out. We went to a party at Princeton, and later to a little sports bar at Rutgers. It did me good to spend some time relaxing with friends to unwind. We were in early, and I was able to get a good night's sleep.

The big day was finally here, and I was ready for action. I was still stinging from the poor practices last week, being kicked out of practice, and being fired as captain. It took me some time to get into the moment and focus on what needed to be done. I made sure I had on my lucky "weak link" t-shirt.

The tournament started with several pools of six fencers, who bouted with the other five fencers in their pool, to bouts of five points. The competitors with the most wins advanced to the next pools of six, until they make the bracket for the double elimination bouts, which go for ten points each.

My fencing was sloppy, and I was out of sorts. I was not doing well. The saving grace was my family and friends who made it a point to cheer me on. Slowly, I started to come alive. I barely made it out of the first pool, and somehow kept advancing to the bracket.

I made it through my first couple of bouts, and started to feel like my old self. Then I faced a fencer from the Naval Academy. He was a nice kid, and I could tell he really wanted this win. Notre Dame fencing is no different from Notre Dame football when it comes to other teams showing up to play. Teams that are mediocre will show up and play as if they are in the Super Bowl™ when they face Notre Dame. This cadet from Annapolis was no exception. He came at me with everything he had. That was exactly what I needed to help me focus and get into a rhythm. I defeated him, but was grateful for his effort.

Later, I was seeded against an Air Force cadet. I had been watching and chatting with him earlier. He was a very solid fencer, probably because the Olympic training center is so close to the Air Force academy. It was obvious this guy had been competing against members of the Olympic team.

As I prepared for my bout, I reflected on various bouting scenarios we practiced. We would set up scenarios like you have 30 seconds left and you are down four to one, and you must execute your plan to win. One of the scenarios was running the clock out. After ten minutes, there would be an added minute. All you had to do was score one point and run out the clock. I decided this was the best course of action for this competitor. I had never used it before in real competition, but my gut told me to roll with it. The bout started and I feigned being aggressive, but I was just biding my time to score that one touch. We went back and forth for what seemed an eternity, and then I saw it. He opened up his arm, and I hit him in the hand. Now the trick was to be aggressive enough to look like I was going to score, but I was really keeping my distance and staying away. *If you cannot reach me, you cannot hit me.* Finally, the ten-minute timer went off. Believe me, that was the longest ten minutes of my life. With one minute to go, I think he realized what I was doing, and he really came after me. With seconds left, he initiated one last attack, and it resulted in a double hit, which meant, we both got a score. I had run out the clock, and won the bout. The crowd went crazy. I was going to the championship bout.

My opponent in the championship bout was from Columbia University, and trained at the very well respected New York Fencing Club. He later went on to join the Olympic team. He had just seen how I beat the Air Force cadet so that strategy was out the window. I would have to try to get in safe hits and keep my distance. I scored first, and we were off to the races. My mother and my Uncle Jeff were cheering so loud that the director had to keep telling them to quiet down because it was not a football game. That brought a smile to my face and helped keep me charged. We went back and forth on the scores, until he finally won 10 to 7. We were given 15 minutes to rest before we started the next bout.

Because it was double elimination, I was going to have to beat him twice. The day had been long, starting with warming up at approximately six in the morning and it was already nearly five in the evening. The only thing I had eaten all day was bananas, apples, and Gatorade™ to keep from dehydrating and cramping. It was starting to take its toll. We started the next bout and we went back and forth for the first five touches. I could feel my energy slipping. I gave it one last push, and it was not enough. I lost 10 to 5. I was depleted, disappointed, and my body was shaking. I was so close. If I had had any more fluid in my body, I probably would have cried. Then the student manager, George, came up to me and told me that my performance had moved the team up to second place. Even if I had won, it would have been mathematically impossible for the team to get anything higher than second place. It was a consolation that the epee squad was no longer the "weak link." We were the "strongest link!"

Life can hit super hard, and it is easy to use excuses to give up. In this case, I had the distractors of unsupportive assistant coaches, feeling I failed myself, feeling I failed my team, feeling I failed Coach DeCicco, feeling frustrated from a terrible week of practice, and feeling low for losing my position as captain. My confidence was shaken to the core. By the grace of God, I was able to find support, keep positive, rely on my training, and dig deep to overcome these obstacles.

Setting your goals high is always a great thing because if you fall short, you still will achieve much more than if you set low goals or no goals. Even though I was stinging from my loss, I felt like I accomplished my obligation to the team and redeemed myself. I was on a mission that day and moved the team up to second place with a 50% win/loss record for the day. The icing on the cake was that I won the First Team All-American Award! I could look back and honestly recognize there was nothing left for me to have changed the outcome. I gave it my best, and left it all on the strip.

Activity

Your activity for this chapter is to write down the equivalent of our bouting scenarios. Look at each possible problem that could arise in your quest, and write down various plans to negotiate the obstacles. Just like football coaches have plays prepared for every scenario, you should also have plays ready to execute your goals. Also, remember when you are knocked down that someone else went through the experience of being knocked down, and with God's help, was successful at getting back up.

The Trophy (courtesy of Leslie Griffee)

21

LIFE ON LIFE'S TERMS

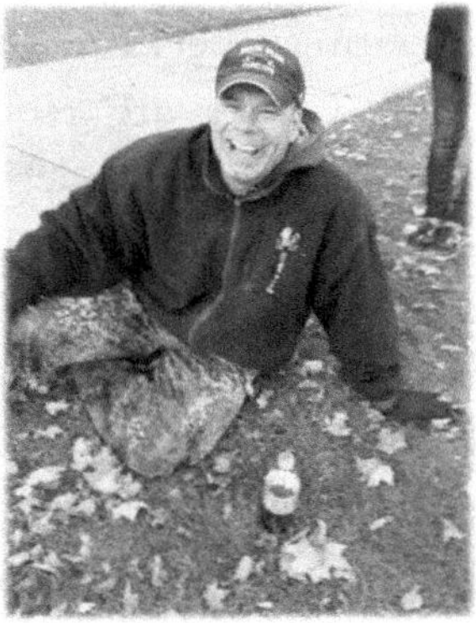

Enjoying campus (courtesy of Bonnie Jo Griffee)

Following the exhausting competition, I went to dinner with my family, then flew home first thing in the morning. The more I thought about it, the angrier I got at the lack of support I received that semester. I was also bitter with the loss of being captain. I fulfilled my debt to the fencing team, so maybe it was time to move on. I got off the plane, did not speak to anyone, and skipped all of the post-season team activities. Coach DeCicco knew something was up and scheduled a meeting with me and my parents. During that meeting, my parents went off about how I had been treated that semester; they told him I was not going to participate my senior year. Coach reassured my parents and myself that would not happen again. He stated he had hired a new assistant epee coach, Adam, a graduate student, who had defected from Poland. Coach stated I would only have contact with himself and Adam. I agreed to stay.

During summer break between junior and senior years, Army ROTC Cadets are required to attend Advanced Camp, which is a leadership style basic training for future officers. My training was at Fort Lewis, Washington. It was beautiful—when it was not raining—which was most of the time. Fire ants built nests three feet high out of pine needles and the mosquitos were from the Mesozoic age. The entire place was covered in volcanic ash from Mount St. Helen's' eruption a couple years prior (in 1980). The training was fun, but demanding.

I suffered some injuries, but the worst was tripping over a log full of yellow jackets. They were trying to crawl in my eyes, ears, mouth, and were stinging me everywhere. They were even inside my uniform stinging me. I crashed into the road, and heard someone ask who was using all that foul language. It was the general and a full bird colonel. Once they noticed I was covered in yellow jackets, they started beating them off me. Fellow soldiers helped me and ripped my clothes off. I was stung everywhere; I was having trouble breathing and started to feel sick. I had to carry my gear two miles to catch a bus to get to medical services to receive a shot.

Despite the strenuous conditions, the training camp was a lot of fun, and a very positive experience. I enjoyed my time in the Army and felt they really did a fine job developing leadership skills which I have used throughout my life.

Now that my Army training was complete, I was glad to be home, getting in the groove, and starting my senior year.

My senior year was by far the most enjoyable. I had completed all of the hard classes and I was taking interesting classes. Even though it was a tough pill to swallow watching someone else lead the epee squad, I was a good scout and did not rock the boat. The first lesson Army ROTC taught us was you had to be a good follower to be a good leader, and I took that lesson to heart.

The season was uneventful. I ran up my win/loss record, and once again double qualified for the NCAA national championships at Northwestern University. I was excited. This was my last chance. I was on fire during practice the week before the competition.

I started the competition miserably. Everything that could possibly go wrong went wrong. I had the worst day of my career. I was eliminated before I could reach All-American honors. I was crushed. Ten years of intense training resulted in this. There was no explanation for the defeat. The chariot wheels got stuck in the mud, and I got wiped out.

A couple of weeks after the competition, Adam and I went to a major tournament in Cleveland, and I did amazingly well. Why couldn't that have happened two weeks prior? I just had to let it go. I had classes, Army, and the rest of my life to focus on. Graduation was right around the corner.

I reflected on my time at Notre Dame. Dr. Koellner was right; nobody could take those experiences from me. Notre Dame is a very special place. It is the place where I learned so many major life lessons, and met so many wonderful friends, professors, teammates, and mentors. I felt I had contributed as much as I received, and was humbly grateful to be a part of the Notre Dame story.

I was commissioned as a second lieutenant in the Army, and graduated the same weekend. Just like a flash in the pan, it was time for me to start my life away from Notre Dame.

Activity

Your activity for this chapter is to reflect on achieving your goal. Make a list of lessons learned so you can use it in future endeavors.

_____

_____

_____

CONCLUSION

Navy Police Commissioning (courtesy of Joe Griffee)

I went on to serve in the Army, and later became a state trooper. I married and had three wonderful kids. As life finds a way, I was knocked down repeatedly. I went through divorce, remarriage, loss of employment, financial difficulties, family deaths, prosecution, election loss, and alcoholism.

During this time, I owned a British pub, and Coach DeCicco would come in from time to time to reminisce about old times. He told me I did not have much natural talent, and was as graceful as a bull in a china shop. He went on to explain that despite those challenges, he had never had a fencer work as hard, and absolutely refuse to lose. He said he respected me for never giving up. That was a valuable compliment from a man who coached so many phenomenal fencers.

Life had its ups and downs and had reached a breaking point where I had a heart-to- heart chat with my father. He advised me to get back into my original career path. I thought to myself, *how do I do that?* I reflected on my life described in the previous 21 chapters, and the instructions were there the entire time. I followed those instructions, and was hired as a federal police officer on a Navy SEAL Base in Virginia Beach. I was later promoted and transferred to a Navy base in Southern Indiana. From there, I was promoted to investigator in the Navy Criminal Investigation Division (CID). I was then promoted to be an NCIS investigator.

Remember at the beginning of the story, when that 10-year-old kid wanted to be a G-man? Well, here I was. The story has come full circle. Of all the places I have worked, the Navy has by far treated me the best. I have been blessed with my career, my beautiful wife, Bonnie Jo, three wonderful children, two great stepchildren, and five awesome grand kids, with one more on the way. God had truly blessed me with a wonderful life, and He was kind enough to give me the instructions on how I accomplished it.

My hope for readers is that all of you have the courage to go after your dreams and passions. Listen to children's answers when adults ask them what they want to be when they grow up. The kids will say ballerina, veterinarian, police officer, firefighter, musician, artist, etc. I wonder how many of those children go on to fulfill those dreams. If they did not, how many brilliant people in those roles and careers did the world miss having? Can you imagine if someone told Yo-Yo Ma he should not be a cellist, Chuck Yeager he could not become a pilot, or Martin Luther King, Jr. he could not become a minister?

Growing up in a tavern, and later owning a pub, I heard countless stories of lost love, lost careers, and lost opportunities—people who had a shot in the major leagues, the NFL, modelling in New York, singing opportunities in Nashville, or their dream corporate job. These stories were heart-breaking, and they all saved face with an excuse of why the opportunity fell through. These broken people were stuck living in the past. They were trying to drown their misery caused by regret.

I am fortunate I do not have to live with the regret of not taking an opportunity. By taking the risk and pursuing your passion, you will always reach a higher goal than if you never attempted the journey. Even if you fail your original goal, do not worry about the odds. I defeated the odds being accepted into Notre Dame, becoming a state trooper, an entrepreneur, and an NCIS investigator. I did not do this because I am perfect. I did this by following the principles of this book, and trusting God to be my partner on the journey.

God always placed the right people in my life at the right time. God placed people in my life who believed in me when no one else did—people like my parents, aunts, uncles, cousins, grandparents, teachers, coaches, Charlie, Fr. Andy, Twisted Tommy, Mike, Alan, and my NCIS mentor, Jim, and those I've named here. I apologize to all those great friends who are not named. There are simply too many.

Like I said before, I am a flawed human being, but God still worked miracles in my life. Have faith that He can work miracles in yours. I encourage you to use these principles so you can be the "strongest link" in your relationships, careers, teams, organizations, and pursuits.

Our Lady of Victory, pray for us!

La Familia (courtesy of Bonnie Jo Griffee)

PRAYER

Our Lady of Victory (courtesy of Todd Griffee)

After years of morning prayers and meditation, I devised this prayer. I took some of it from various sources, but it works for me. Feel free to write your own prayer to facilitate your daily dialogue with God:

God, in the name of Jesus Christ your Son, please send your Holy Spirit on me to remove all cancer, and this fat. Please remove from me all sickness, illness, injury, and disease, [list additional ailments here]. Please remove from me all mental blocks, and musical blocks. Remove from me all jello-brain, forgetfulness, and fatigue. Remove from me all self-centered, self-seeking, and selfish behavior. Remove from me all anger, fear, hate, and resentment. Please remove from me all financial worries, relationship issues, [list your own intentions here]. Breathe in me your healing spirit, and fill me with a clean spirit, a right heart, an able body, and a sound mind. Fill me with faith, love, hope, and charity.

Thank you, God, for this beautiful day and the gift of sobriety. Without that gift, I would not have the people, places, things, experiences, and opportunities in my life today.

Blessed Mother Mary, please pray with me for my family, friends, coworkers, and their families, [insert names of specific people]. Please pray for all who need prayers, all who have passed, all I have hurt, all who have hurt me, all I could be resentful of, and protection from my enemies.

Say a Hail Mary, a Glory Be and an Our Father.

God, please take all of me, good and bad. Please remove from me any defect of character within myself or obstacle in the outside world which would get in the way of Your will and Your life for me today. That victory over my problems will provide hope to others, because if you can work miracles in my life, you can work miracles in theirs. Please do not let me say or do anything stupid today. Please let me do good work, and get through this day.

Touchdown Jesus (courtesy of Bonnie Jo Griffee)

About
Kharis Publishing:

Kharis Publishing, an imprint of Kharis Media LLC, is a leading Christian and inspirational book publisher based in Aurora, Chicago metropolitan area, Illinois. Kharis' dual mission is to give voice to under-represented writers (including women and first-time authors) and equip orphans in developing countries with literacy tools. That is why, for each book sold, the publisher channels some of the proceeds into providing books and computers for orphanages in developing countries so that these kids may learn to read, dream, and grow. For a limited time, Kharis Publishing is accepting unsolicited queries for nonfiction (Christian, self-help, memoirs, business, health and wellness) from qualified leaders, professionals, pastors, and ministers. Learn more at: https://kharispublishing.com/

Printed in the USA
CPSIA information can be obtained
at www.ICGtesting.com
LVHW022117071024
793187LV00008B/269